# An Introduction to Physics and Technology of Thin Films

# An Introduction to Physics and Technology of Thin Films

**Alfred Wagendristel**
Institute of Applied and Technical Physics
Technical University of Vienna

**Yuming Wang**
Department of Materials Science
Jilin University

*Published by*

World Scientific Publishing Co. Pte. Ltd.

P O Box 128, Farrer Road, Singapore 9128

*USA office:* Suite 1B, 1060 Main Street, River Edge, NJ 07661

*UK office:* 73 Lynton Mead, Totteridge, London N20 8DH

**Library of Congress Cataloging-in-Publication Data**

Wagendristel, Alfred.
An introduction to physics and technology of thin films / Alfred Wagendristel, Yuming Wang.
p. cm.
Includes bibliographical references and index.
ISBN 9810216165
1. Thin films. 2. Thin film devices. I. Wang, Yu-ming. II. Title.
QC176.83.W34 1994
530.4'175--dc20 94-13920
CIP

Printed in Singapore.

# INTRODUCTION

A thin film is a liquid or solid such that one of its linear dimensions is very small in comparison with the other two dimensions. Usually one classifies thin films (arbitrarily) into:

thick films ($D > 1$ micrometer, $D$ : film thickness)
thin films ($D < 1$ micrometer).

This book discusses mainly the systems of a (solid) film on a (solid) substrate (backed films) rather than unsupported films (foils). They (film + substrate) imply a production process in the form of film growth by sustaining an atomic or molecular flux to the surface of the substrate and subsequently by growing of the film. Film growth will either involve chemical reaction at the substrate such as discharge of ions, decomposition of a compound, reaction of a gas or liquid with substrate surface; or physical processes such as evaporation from a source and sputtering from a target, then condensation onto the substrate.

The history of thin film technologies is as follows:

1650 Interference colors of thin liquid film on a liquid surface (oil on water) were observed by R. Boyle, R. Hooke, and I. Newton.

1850 Electrodeposition (by M. Faraday), Chemical reduction deposition, Film formation during glow discharge (by W. Grove), and Evaporation of metallic wires by current (by T. A. Edison) were discovered.

Solid films produced by the first two methods received early recognition for their technical importance as anticorrosive films or films for mirrors, whereas those prepared by the latter methods lacked reproducibility for a long time. Only since the improvement of vacuum equipment for film preparation as well as for investigation (electron microscopy, LEED, other surface analytical techniques) were reproducible and useful films readily obtained. Since 1950 a vigorous development arose by production of

1940 Films for optical, electronic, mechanical, and protective applications.

1965 Semiconductor electronics made use of thin film methods which reveal two major merits: mass fabrication by printing techniques, and miniaturisation by integration (integration density: 1966: 50 elements/mm$^2$; 1974: 500 elements/mm$^2$; 1980: 5000 elements/mm$^2$)
Main fields of application:
computer electronics, commercial electronics, medical electronics, space technology, and energy conversion (solar cells).

Optical applications of metal and dielectric films:
filters, reflection coatings, optical wave guides for opto-electronic communications of semiconducting films:
IR-sensors and thin film laser diodes.
Magnetic and superconducting films for memory and logical devices.

Basic scientific interest is focussed on film formation processes in order to obtain an insight into the mechanisms leading to special structural properties:

Island-, labyrinth-, continuous films (macrostructure) amorphous, polycrystalline, single crystal films (microstructure) on film stability (the mobility of atoms in the films enable relaxation and heterodiffusion).

It is, however, not only the film structure but also the limited thickness which determines the physical properties. Thus the geometric anisotropy and size effects should also be studied.

In the following chapters, firstly methods of film deposition with special emphasis on the basic physical phenomena, then the fundamental processes in film formation, and finally some thin film specific properties and their technical relevance will be discussed.

# CONTENTS

# PART I

# CHAPTER 1

# THIN FILM DEPOSITION METHODS

## 1.1. Chemical Methods[1]

### 1.1.1. *Electroplating*

This method is suitable only for deposition of metals and alloys on electrically-conducting substrates. The film material is present in the form of positive ions in the electrolyte (mostly an aqueous solution of an ionic compound). The number of ions discharged at the cathode and hence the mass of deposit is given by Faraday's law.

$$\frac{m}{A} = \frac{jtM\alpha}{nF} \tag{1}$$

where $m/A$ denotes mass per area, $j$ current density, time, $M$ = molecular weight, $n$ = valency, $F$ = 96490 As/g equivalent, $\alpha$ = current efficiency = 1 − 0.5. Among the 70 metallic elements there are 33 which can be electrodeposited but only 14 of them are deposited routinely:

Al, Ag, Au, Cd, Co, Cu, Cr, Fe, Ni, Pb, Pt, Rh, Sn, Zn.

For physicists, the elementary process of discharge and film formation is of interest: Ions (of both types) are accelerated towards the oppositely-charged electrode, where they form a double layer screening the bulk electrolyte from the main part of the electric field. The voltage drops in the double layers (about 30 nm thick) leading to fairly high field strengths

($10^7$ V cm$^{-1}$). In an aqueous system the positive ions undergo several reactions before being incorporated in the films:

1) dehydration
2) discharge
3) surface diffusion
4) nucleation, crystallisation.

In general polycrystalline deposits but in special cases epitaxial growth on a single crystal cathode is possible, e.g., Ni on Cu.

For alloy deposition one must account for different electrochemical potentials governing the ratio of discharged ions of the components at a certain voltage. Suitable chemical complexing of ions may adjust the particular discharge reactions. Special feature of electrodeposition is the high growth rates $\dot{D} = dD/dt = 1\ \mu\text{s}^{-1}$ at a current density $j = 1$ A cm$^{-2}$.

### 1.1.2. *Electroless plating (chemical reduction plating)*

In some special cases electrochemical reactions may occur without an external field, e.g., silvering of mirrors by $AgNO_3$ solution with formaldehyde or sugar as a mild reducing agent. This reaction takes place at any surface submersed in the bath. Sometimes the reaction will be realized only on special surfaces:

NiCl + sodium hypophosphite $\rightarrow$ Ni deposit only on Ni, Co, Fe and Al surfaces. It is possible to activate surface parts with special substances, e.g., $PdCl_2$ solution is used for activating nonmetallic surfaces for Cu or Ni deposits, $SnCl_2$ for activating nonmetallic surfaces for Ni films.

Special features of this method are: highly specific, difficult to control but suitable for nonconducting substances and for deposition on places difficult to be reached, e.g., inside Dewar vessels.

### 1.1.3. *Chemical vapor deposition (CVD)*

Such methods that are making use of a gas transport reaction are very important in semiconductor device fabrication.

The film material is usually one of the components in a volatile compound which decomposes at the substrate in the form of a heterogeneous reaction. Elevated temperatures are usually required, thus enabling the growth of single crystal films on a suitable substrate, very often iso-epitaxial growth of Si on Si single crystal.

Table. 1. Summary of materials that can be deposited by vapor-phase reactions (parent material in parentheses)

| Method | Material |
|---|---|
| Disproportionation ($A+AB \leftrightarrow 2AB$) | $Al(AlI_3)$; $Ge(GeI_2)$; $Si(SiI_2)$; III–V compounds(iodides); $C(CO)$ |
| Polymerization ($AB \rightarrow nAB$) | Polymers of methyl methacrylate, styrene, divinylbenzene, butadiene, acrolein, epoxy resins, allylglycidyl ether, etc. (by electron-beam, photolysis or glow-discharge) |
| Reduction $H_2$ ($AX \rightarrow A+HX$) | $Al(AlCl_3)$; $Ti(TiBr_4)$; $Sn(SnCl_4)$; $Ta(TaCl_5)$; $Nb(NbCl_5)$; $Cr(CrCl_2)$; $Si(SiHCl_2$ or $SiCl_4)$; $Ge(GeCl_4)$ |
| Oxidation $H_2O$ ($AX \rightarrow AO+HX$) | $Al_2O_3(AlCl_3)$; $TiO_2(TiCl_4)$; $Ta_2O_5(TaCl_5)$; $SnO_2(SnCl_4)$ |
| Nitriding, etc. | $Si_3N_4$ (from $SiH_4$ by pyrolysis; glow discharge); |
| $NH_3$ ($AX \rightarrow AN+HX$) | $TiN(TiCl_4)$; $TaN(TaCl_5)$; $SiC(SiCl_4+CH_4)$ |
| Decomposition ($AB \rightarrow A+B$) | $SiO_2$ (from Si esters by pyrolysis or glow discharge); $Ti(TiI_4)$; Pb(Pb-organics); $Mo(MoCl_5)$; $Fe[Fe(CO)_5]$; $Ni[Ni(CO)_4]$; C(toluene); $Si(SiH_4)$; $MnO_2[Mn(No_3)_2]$; BN(B trichloroborazole) |

Some of the reactions that are used practically are listed in Table 1.

In Fig. 1, a closed system for Si and Ge using the disproportion of their iodides is shown. Other important reactions for semiconductor fabrication are the pyrolysis of $SiH_4$ and the reduction of $SiCl_4$ or $SiHCl_3$ with H. These methods however require a somewhat more complex setup.

Special features are: High temperature is usually required, epitaxy is achieved, rather high growth rate is possible, material difficult to evaporate can be deposited. Chemical vapor deposition, however, is experimentally complex and sophisticated.

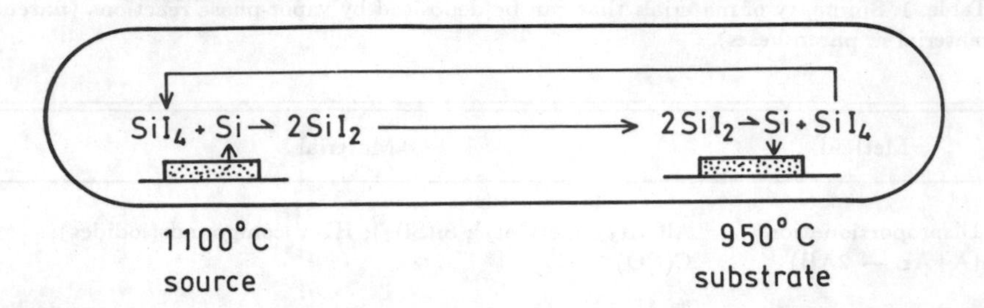


Fig. 1. A CVD closed system for Si film using the disproportion of its iodides.

### 1.1.4. *Hydrophily (Langmuir–Blodget films)*

Certain long-chained fatty acids and their salts show an atomic structure such that one end is hydrophilic (e.g., Ba-stearate). Such molecules swim in an "upright" position on water surface. If a proper solution of such molecules in a volatile solvent is spread on water surface, monomolecular layers of the fatty acid will cover the surface from which they can be fished up by a substrate as shown in Fig. 2. By repeating this procedure rather thick insulating films can be made which, if sandwiched by metal films, may form capacitors or tunnelling devices.

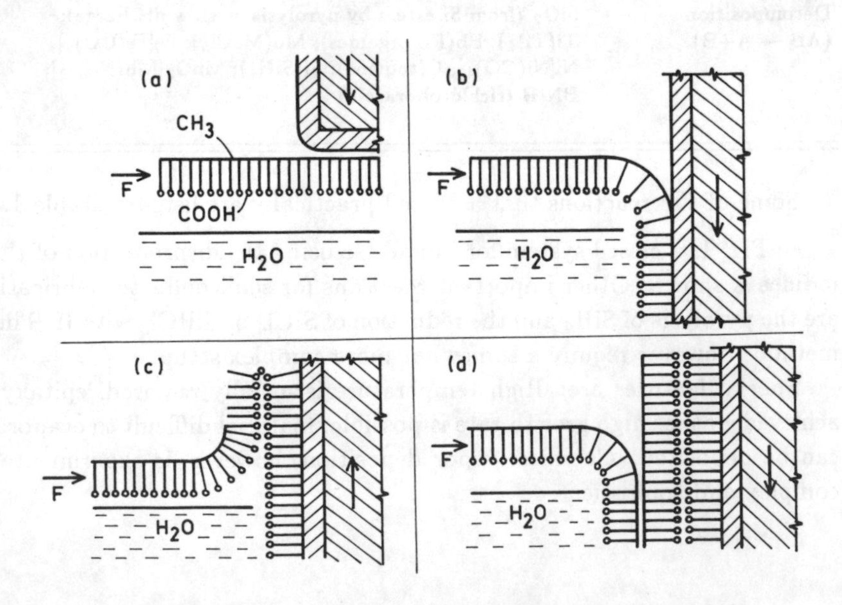


Fig. 2. Process of Langmuir–Blodget film production.

### 1.1.5. *Anodization*

The reacting partner is atomic oxygen and hence a very reactive state is produced by the decomposition of $H_2O$ next to the anode of an aqueous electrolytic system. The reactions are

$$\text{a: } M + nH_2O \rightarrow MO_n + 2nH^+ + 2ne^-$$
$$\text{c: } 2ne^- + 2nH_2O \rightarrow nH_2 \uparrow + 2nOH^-$$

The growing films are insulating, thus self-limiting their growth process. Well known examples are $Al_2O_3$ films grown on Al for protection, decoration and for electrolytic capacitors. Anodic films are generally amorphous. The thickness limit $D_{max}$ depends on the voltage $U$ applied: $D_{max} = KU$, where $K$ is a material coefficient. $D_{max}$ and $K$ for various materials are listed in Table 2.

Table 2. Thickness limit $D_{max}$ and coefficient for several elements

| Element | Al | Ta | Nb | Ti | Zr | Si |
|---|---|---|---|---|---|---|
| $K$ (A/V) | 3.5 | 16.0 | 43.0 | 15.0 | 12.0–30.0 | 3.5 |
| $D_{max}$ ($\mu$m) | 1.5 | 1.1 | | | 1.0 | 0.12 |

### 1.1.6. *Thermal growth*

It is known that a reactive metal surface such as that of Pb and Al already forms an oxide layer at room temperature when exposed to atmosphere. This process is a very important one in semiconductor technology where Si is oxidized in water vapor at high temperatures (about 1000°C). The $SiO_2$ thus formed has excellent insulating and protecting properties. Nitridization is possible as well if $NH_3$ is the reactive gas. Temperatures can be generally reduced if a reactive plasma is created and the ions are accelerated towards the substrate. The growth relation is a parabolic one since O has to diffuse through the layer already formed. Thickness is limited by the internal stress which, if exceeding a critical value, makes the films partially split off the substrate. The thickness $D$ growing with time $t$ at various $T$ is shown in Fig. 3.

**Table 3. Summary of ways of preparing thin films**

| Method | Deposition Rate, Ås$^{-1}$ | Rate Control | Type* | Advantages | Limitations |
|---|---|---|---|---|---|
| Electroplating | $10^2$–$10^4$ | Current density | M | Simple apparatus | Metallic substrate |
| Chemical reduction | 10 | Solution temp., pH | M | Simple apparatus | Limited number of materials |
| Vapor phase | 1–$10^3$ | Pressure, temp. | M,S,I | Single crystal, clean films | High substrate temp., low pressure |
| Anodization | 10 | Current density | I | Simple apparatus, thin amorphous films | Metallic substrate, limited number of metals, limited thickness |
| Thermal | 1 | Pressure, temp. | I | Simple apparatus | Metallic substrate, limited thickness, limited number of metals |
| Evaporation | 10–$10^3$ | Source temp. | M,S,I | Large range of materials and substrates | Vacuum apparatus, some materials decompose on heating |
| Sputtering | 10 | Current, potential | M,S,I | High adhesion, very large range of materials | Suitable targets, vacuum apparatus |

*M – Metals; S – Semiconductors; I – Insulators.

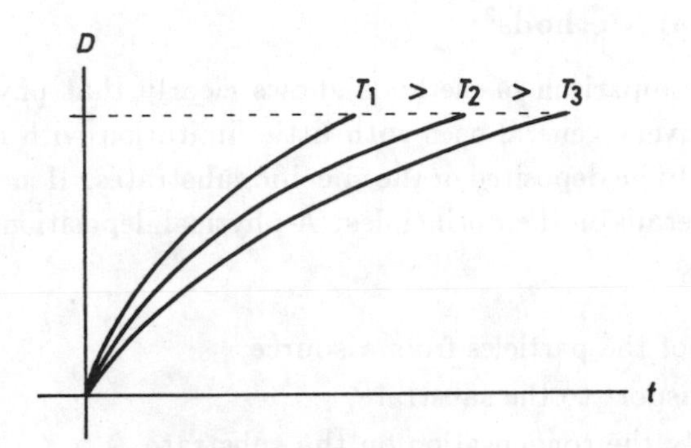


**Fig. 3. Film thickness $D$ vs time $t$ during thermal growth at various temperatures $T$.**

## 1.1.7. *Summary*

The features, merits and limitations of the methods discussed are outlined in Table 3. Table 4 shows which method is suitable for the production of films of special material.

**Table 4. Deposition system used for preparation of films and devices**

| Film Type | Device | System | | | | | | | |
|---|---|---|---|---|---|---|---|---|---|
| | | Evap | Sputt | Thermal | Anod | VP | EP | CRP | Sol |
| Metallic | Conductors | X | X | | | X | X | | |
| | Resistors | X | X | | | | | X | |
| | Magnetics | X | X | | | | | X | X |
| | Superconductors | X | | | | | | | |
| Dielectric | Capacitors | X | | | X | | | | |
| | Insulators | X | X | X | X | X | | X | |
| | Optics | X | X | | | | | | |
| | Ferroelectrics | X | | | | | | | |
| Metal/ | Resistors | X | X | | | X | | | |
| Dielectric mixture | Optics | X | X | | | | | | |
| Semiconducting | Resistors | | | | | X | | | |
| | Active | X | X | | | X | | | X |
| | Photoconducting | X | | | | X | | | |
| | Photovoltaic | X | | | | X | | | |

## 1.2. Physical Methods[2]

The above comparison of methods shows clearly that physical deposition methods are very general ones with little limitation with respect to either the material to be deposited or the specific substrates. Hence we are putting some more details on their principles. A physical deposition process consists of three steps:

emission of the particles from a source;
their transport to the substrate;
and finally the condensation on the substrate.

In the same sequence the details of physical vapor deposition (PVD) will be discussed as follows:

### 1.2.1. *Sputter deposition*

In 1852, Sir W. R. Grove discovered surface coatings generated in the valve where he investigated a glow discharge. Sir W. Thomson called this phenomenon, in analogy to the generation of drops out of a liquid surface by an impinging primary drop, SPLUTTERING. The expression SPUTTERING as a result of a printer's error was soon adopted as a scientific term.

Compared to evaporation, sputter deposition was fairly late on developing up to the modern standard which now offers a vast variety of applications.

#### 1.2.1.1. *Physical fundamentals*

The ejection of particles from a solid surface after exposure to a bombardment with heavy particles (usually ions) of sufficient energy, i.e., the sputtering phenomenon was first believed to result from local heating of the target. Fundamental differences thermal evaporation, however, soon lead to the conclusion that sputtering is the result of a momentum transfer from the bombarding particles. The following experimental facts will make this clear:

1) the angular distribution of sputtered particles depends on the direction of impinging particles (Fig. 4a).
2) particles sputtered from single crystal targets show preferred directions (Fig. 4b).

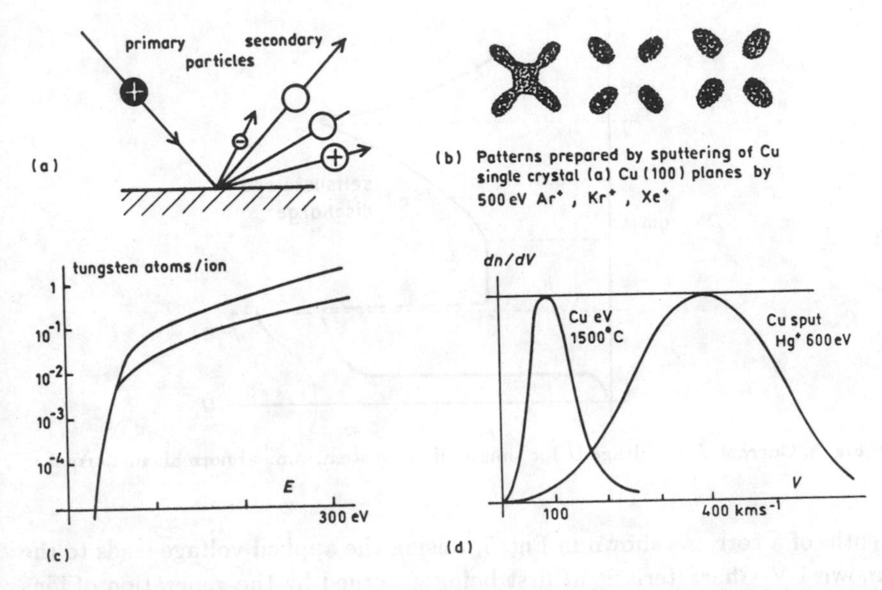


**Fig. 4. Characteristics of sputtering a) Angular distribution of sputtered particles. b) Patterns sputtered from (100) planes of single crystal Cu by ions $Ar^+$, $Kr^+$ and $Xe^+$ of 500 eV. c) Sputter yield vs ion energy. d) Velocity ($v$) distribution of particle number ($n$) during sputtering (sput) and evaporation (ev) for Cu films.**

3) sputtering yields not only depend on the particle energy but also on their mass (Fig. 4c).
4) the mean velocity of sputtered particles is much higher than those evaporating thermally (Fig. 4d).

The phenomena 1, 3 and 4 are plausible if a momentum transfer mechanism is assumed. The preferred directions for single crystal targets can be understood by a focused collision: The impinging ions usually penetrate some atomic radii into the target where they lose their momentum step by step. In special directions the consecutive knocking-on of atoms can lead to a special final direction where the ratio between atomic diameters and distances is within a special range (usually along densely packed directions). An energy balance shows that about 95% of the kinetic energy of the ions is used up for stimulating lattice vibrations in the target (heat) and only 5% is transferred to the sputtered particles.

How to create ions for sputtering? The simplest source of ions is the plasma of a DC glow discharge whose basic details are briefly reviewed as follows. We consider a simple diode system at a gas pressure of several

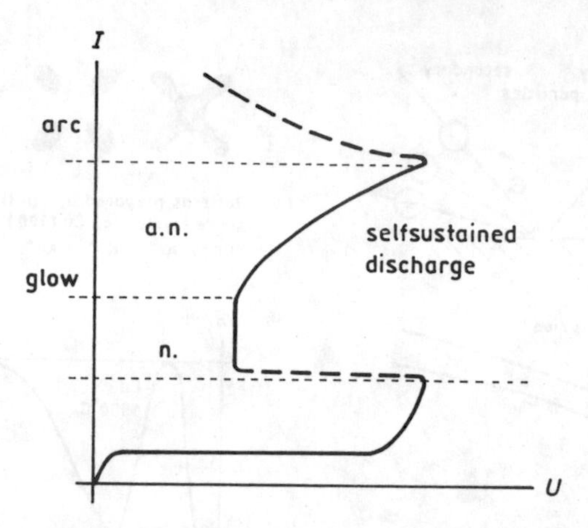


Fig. 5. Current $I$ vs voltage $U$ for simple diode system. a.n.–abnormal. n.–normal.

tenths of a torr. As shown in Fig. 5, raising the applied voltage leads to the known I-V-Characteristic at first being governed by the generation of ions by external causes (radiation). After a certain field strength the primary ions can ionize other neutrals thus creating auxiliary charge carriers for a current. When a critical number is exceeded no further externally created ions are necessary for the discharge. A self-sustained glow discharge is ignited. In the region of normal glow only the discharge's cross section is enhanced for achieving a higher current, whereas in the region of abnormal glow a higher current is associated with a higher energy of ions and hence with a higher voltage. It is this region, where the current can be controlled by the voltage, which is generally used for sputtering. We consider the discharge under steady state conditions as shown in Fig. 6. Because of their higher inertia the ions generate a much thicker double layer at the cathode than the electron layer next to the anode. The major voltage drop, known as the cathode fall, extends over this region, where ions obtain their energy for subsequent sputtering and for the ejection of secondary electrons for sustaining the discharge. The electrons are accelerated towards the anode, first exciting some light emission and finally, when their energy is sufficient for ionization, they ionize at the end of the cathode fall, characterized by a dark room (Krookes DR). Here they lose their energy, thus gathering in a somewhat larger density, manifested in an opposite field. In the region of the positive column, the field strength is low, and charge carriers move by diffusion rather than by drifting. A little drift is only used for exciting light.

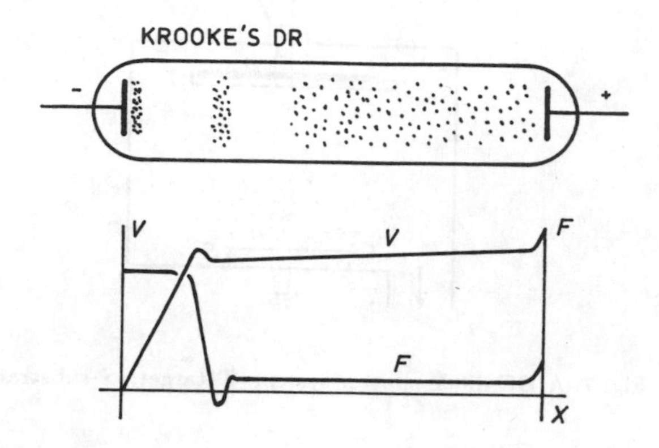


Fig. 6. Voltage ($V$) and field ($F$) distribution along the Krookes tube.

The extension of Krookes DR corresponds mainly to the distance needed by an electron for an ionizing collision. Hence between all environment at anode potential nearer to the cathode than this distance a discharge cannot be ignited. This fact is important for the technical construction of sputter systems. It also fixes the minimum distance between anode (substrate) and cathode (target). The extension of Krookes DR is given by Paschen's law:

$$d = \frac{1}{p}\left(A + \frac{B}{U - C}\right), \tag{2}$$

where $P$ denotes pressure, $U$ voltage, $A, B, C$ constants.

1.2.1.2. *Technical aspects*

Typical data for a DC diode sputter system shown in Fig. 7 are: Voltage $> 500$ V, mostly 1–5 kV, current density = 1–10 mA cm$^{-2}$, Target area = 10 cm$^2$, distance between target and substrate = 10 cm, and the most frequently used sputter gas is argon because of its inertness and low price.

Special techniques can be employed for improving purity of the films by reducing the impingement of gas atoms or by simultaneously removing them from the substrate.

BIAS SYSTEM (Fig. 8)

The substrate is biased negatively (about 100 V) with respect to the plasma (= anode potential), thus exposing it frequently to a tender bom-

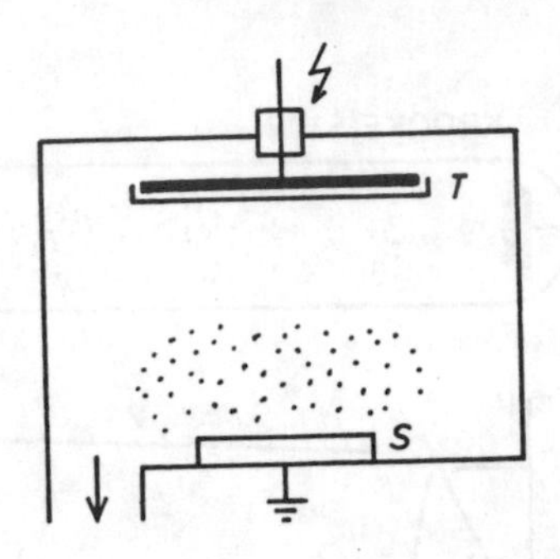


Fig. 7. A DC diode sputter system. $T$–target. $S$–substrate.

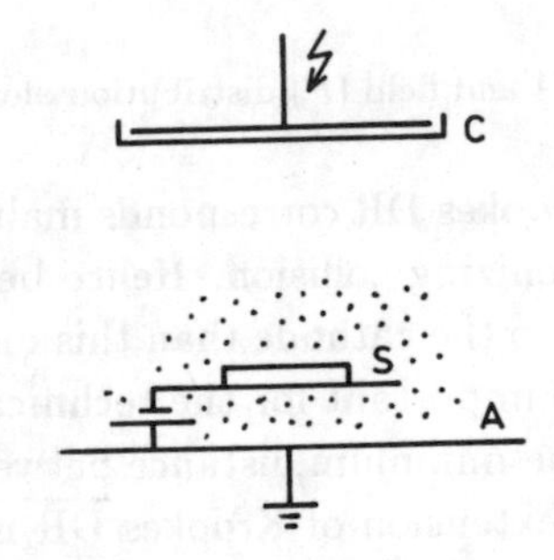


Fig. 8. A DC bias sputter system. C–cathode. A–anode.

bardment with ions knocking out the lighter impurities (trapped gas) rather than the film atoms.

An asymmetrical alternating voltage gives similar results in a step-wise manner. (Fig. 9)

## LOW PRESSURE SYSTEMS (Fig. 10)

Triode sputtering makes use of a thermionic electron source for auxiliary ionization at low pressure. The discharge can be sustained down to about 1 m Torr. Magnetic field may elongate the trajectories of the electrons thus effectively reducing their mean free path (MFP).

DC-sputtering can be used only for sputtering conductors. By ion bombardment insulators would quickly build up a positive surface charge such that no further ions will impinge. The method described in the following section excludes this disadvantage, simultaneously offering the possibility of a further reduction of the pressure.

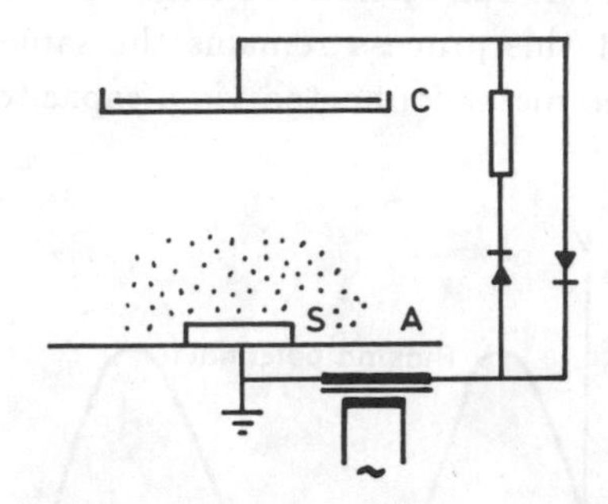


**Fig. 9. A DC sputter system with an asymmetric alternating voltage.**

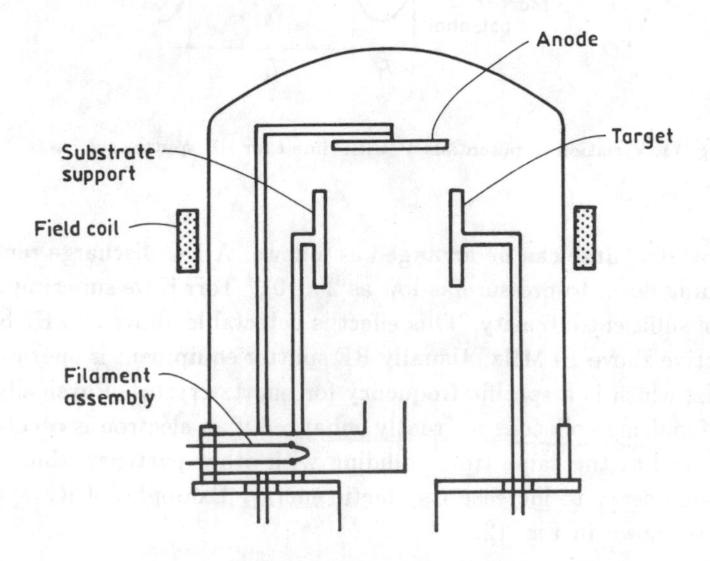


**Fig. 10. Low pressure sputter system with a magnetic field coil.**

## RF SYSTEMS

Principle of sputtering insulators is as follows: A metal part inserted into the plasma and connected to an AC source by a capacitor will charge up negatively, because it draws out more electrons than the heavier ions out of plasma. The equilibrium charge is given by that value which finally extracts the same amount of ions and electrons, i.e., the period being positive against the plasma is much shorter than the negative one. (Fig. 11) Hence by means of this rectifying process, the target is periodically bombarded by ions, while in the complementary period the positive charge build-up is neutralized. It is easy to imagine that this process remains the same if we substitute a dielectric surface for the metal connected via a capacitor.

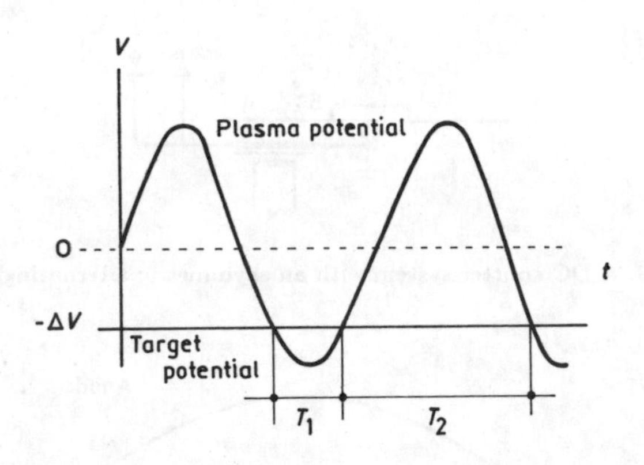


**Fig. 11. Variation of potentials *V* with time *t* for RF sputter system.**

RF glow discharge can be arranged as follows: A DC discharge remains selfsustaining down to pressure as low as $2 \times 10^{-4}$ Torr if we superimpose a RF field of sufficient intensity. This effect is detectable above 50 kHz but is most effective above 10 MHz. Usually RF sputter equipment is operated at 13.65 MHz, which is a specific frequency for quartz crystal. Physically the number of ionizing collisions is greatly enhanced if an electron is oscillating in an RF field at the same time colliding with other particles, thus using the RF field energy to increase its kinetic energy. Examples of RF sputter systems are shown in Fig. 12.

## MAGNETICALLY ASSISTED SPUTTER SYSTEMS (MAGNETRON SPUTTERING) (Fig. 13)

By means of permanent or electromagnets the plasma is confined next to the cathode of a DC sputtering system. The enclosure of the plasma increases the efficiency of a sputter system as the number of atoms available for bombarding the target is highly increased. The plasma is confined to a ring next to the pole piece of the magnet hence sputtering preferably happens at this target region. (Fig. 13) Growth rate up to 10 $nms^{-1}$ are available.

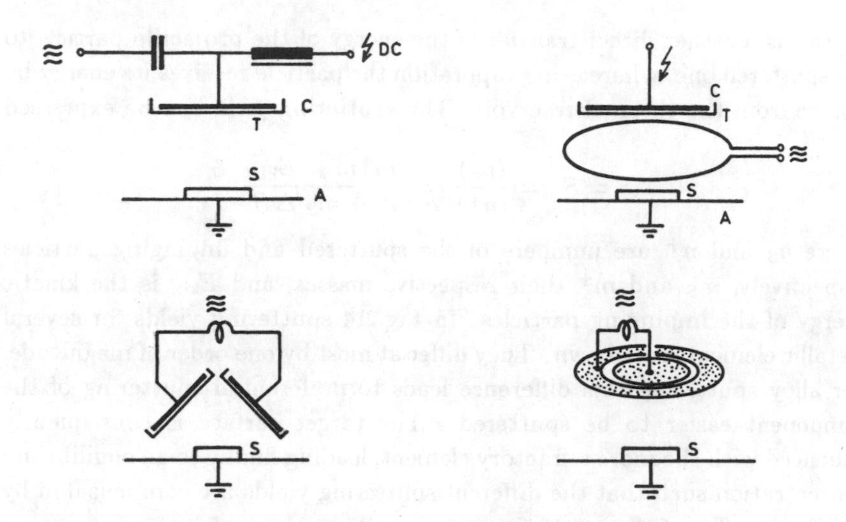


Fig. 12. Different arrangements of RF sputter systems.

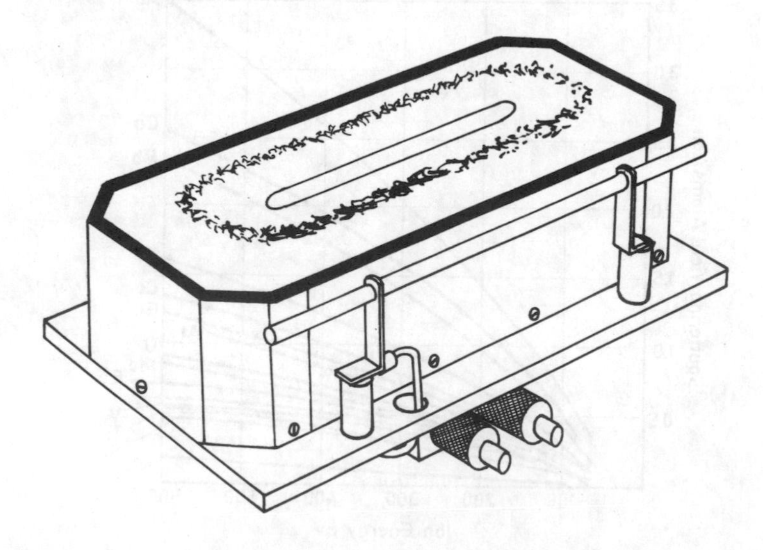

Fig. 13. Cathode of magnetron sputter system.

### 1.2.1.3. *Sputtering yields*

Contrary to evaporation, where the heat of sublimation $\Delta H$ of the source material appears exponential in the number of ejected particles, a much weaker dependence of sputtering yields upon $\Delta H$ is observed. The reason for this is a rather direct transfer of the energy of the projectile particle to the sputtered one, whereas in evaporation the particle receives its energy by chance from the thermal reservoir. The sputtering yield can be expressed as

$$S = S \quad = \frac{\langle n_S \rangle}{\langle n^+ \rangle} = \frac{m^+ m_S}{m^+ + m_S} \frac{E_{kin}}{\Delta H}$$

where $n_S$ and $n^+$ are numbers of the sputtered and impinging particles respectively, $m_S$ and $m^+$ their respective masses, and $E_{kin}$ is the kinetic energy of the impinging particles. In Fig. 14 sputtering yields for several metallic elements are shown. They differ at most by one order of magnitude. For alloy sputtering this difference leads to preferential sputtering of the component easier to be sputtered. The target surface is consequently enhanced with the more refractory element, leading finally to an equilibrium concentration such that the different sputtering yields are compensated by a different offer of the respective components at the surface.

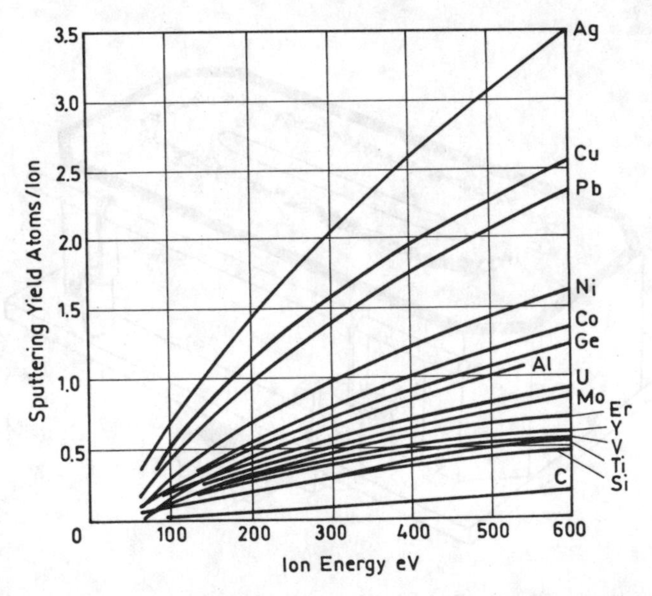


Fig. 14. Sputter yield vs Ar ion energy when sputtering various elements.

The merits of sputter deposition are:

1) Multicomponent films, insulators as well as refractory materials can be deposited.
2) Good adhesion is assured.

3) Thickness uniformity over large areas can be obtained.
4) Thickness control is easy since the thickness is proportional to the deposition time.
5) Substrate cleaning *in situ* is possible by ion bombardment.

The disadvantages are:

1) Source material must be available in sheet form.
2) Deposition rate usually lower than 40 Å $s^{-1}$.
3) Substrate must be cooled except for short runs.

### 1.2.2. *Evaporation*

#### 1.2.2.1. *Physical fundamentals*

The amount of gas atoms at a pressure $P$ impinging on a surface per time $t$ and area $A$ is given from kinetic theory of gases as

$$\frac{dZ}{dAdt} = \frac{P}{\sqrt{2\pi mkT}} \tag{3}$$

where $T$ denotes the absolute temperature, $m$ the mass of the particles and $k$ the Boltzmann constant. In a homogeneous system consisting of vapor and one of its condensed phases in thermal equilibrium, i.e., at vapor pressure $P^*(T)$ the number of evaporating atoms is the same as that condensating. The latter is the amount of impinging atoms at the condensated surface. Hence we can take the above expression for the number of evaporating atoms as well:

$$\frac{dZ}{dAdt} = \frac{P^*}{\sqrt{2\pi mkT}} \tag{4}$$

which is assumed to hold even under non-equilibrium conditions. Then it describes the number of atoms evaporating free in vacuum or in a medium where its partial pressure is zero.

Vapor pressure and temperature are associated in the vapor pressure curve $P^*(T)$ as shown schematically in Fig. 15 whose differential equation is the Clausius–Clapeyron relationship:

$$\frac{dP^*}{dT} = \frac{1}{T}\frac{\Delta H(T)}{V_g(T) - V_c(T)} \tag{5}$$

where $\Delta H$ is the molar heat of evaporation or sublimation, $V_g$ and $V_c$ the molar volumes of the gas and the condensed phase respectively. It can be

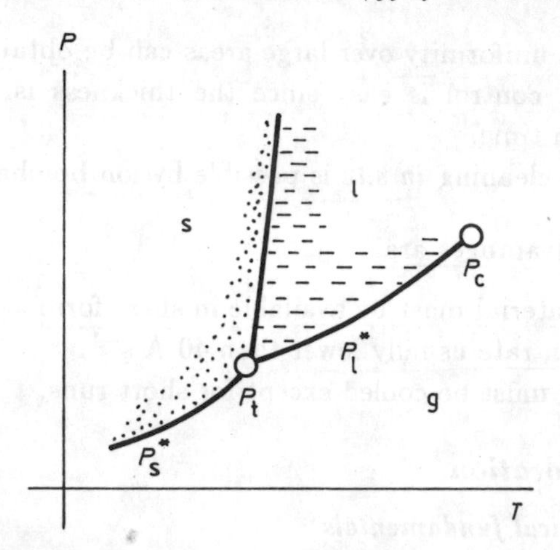


Fig. 15. Pressure ($P$)–temperature ($T$) diagram. s–solid. l–liquid. g–gas.

integrated if $H(T)$, $V_g(T)$ and $V_c(T)$ are known. An approximate solution is obtained for $\Delta H$ = constant and $V_g = RT/P^* \gg V_c$:

$$\frac{dP^*}{P^*} = \frac{1}{R}\frac{\Delta H}{T^2}dT \rightarrow P^* = P_0 e^{-\Delta H/RT} = P_0 e^{-\frac{\Delta H/L}{kT}} \tag{6}$$

where evaporation is considered to be a thermally activated process with an atomic activation energy $\Delta H/L$, $L$ being the Avogadro number. The exponential is due to the probability for an atom to surmount the potential $\Delta H/L$ by receiving adequate energy from the thermal reservoir. Hence we notice that, contrary to sputtering, the number of evaporated particles is much more strongly dependent on the heat of sublimation, i.e., the binding energy of a surface atom.

Equations (4) and (6) show that an appreciable number of evaporating atoms is observed only at higher temperatures. For a practical growth rate of film, $P^* \approx 0.1$ Torr is necessary. The respective temperatures are listed for a few examples:

$$P^* = 0.1 \text{ Torr} \quad \text{Fe} = 1920 \text{ K}$$
$$\text{Cu} = 1790 \text{ K}$$
$$\text{Al} = 1640 \text{ K}$$
$$\text{Sb} = 885 \text{ K}$$

They can be obtained for a variety of metals from the vapor pressure curves depicted in Fig. 16.

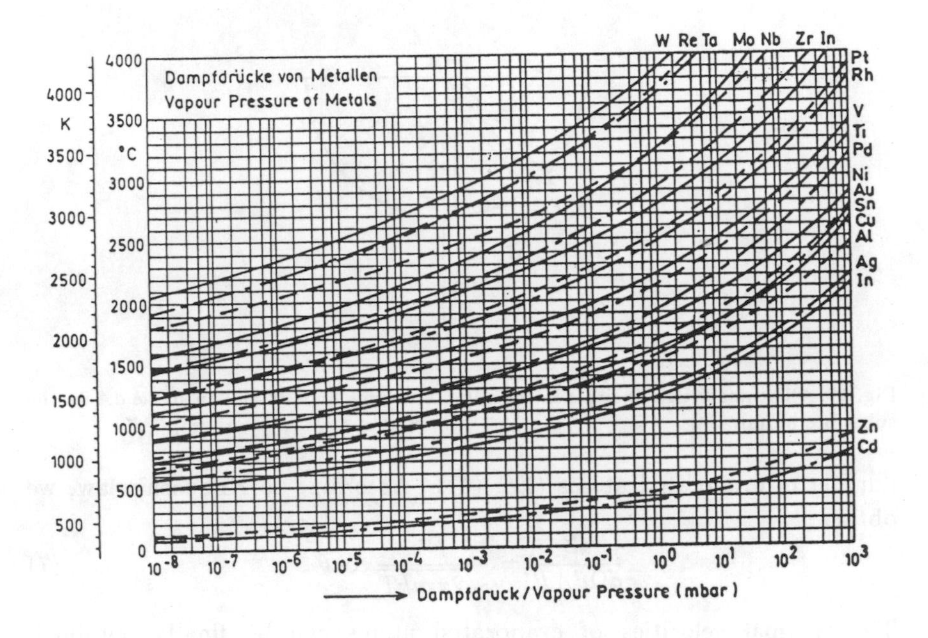


**Fig. 16. Vapor pressure of elements at various temperatures.**

In the above considerations we assumed that evaporation occurs at the surface only. Why don't we expect bubble formation similar to boiling water? One can easily estimate that a temperature gradient suitable for creating a vapor pressure equal to the hydrostatic pressure usually cannot be sustained because of the high thermal conductivity of the source material. Hence for metals the above statements are correct. For poorer thermal conductors such as many dielectrics one in fact sometimes obtains the splitting of larger aggregates by exploding bubbles next to the surface.

For positioning of the substrates, the vapor density as a function of the direction with respect to the surface normal must be known. One can start once again with a vapor in equilibrium with its condensed phase. Then, not only the total numbers of condensing and evaporating atoms are equal but also the densities in each element of the solid angle must be equal to each other. Hence we may say that the angular distribution of the velocities of atoms is isotropic in the case of equilibrium and the number of those striking unit area per second is proportional to the number of atoms in a tube $cn \cos\theta \, dA$, $n$ being the number of atoms per unit volume and $c$ their velocity, as shown in Fig. 17. Integration over $\theta$ and $c$ must yield the total

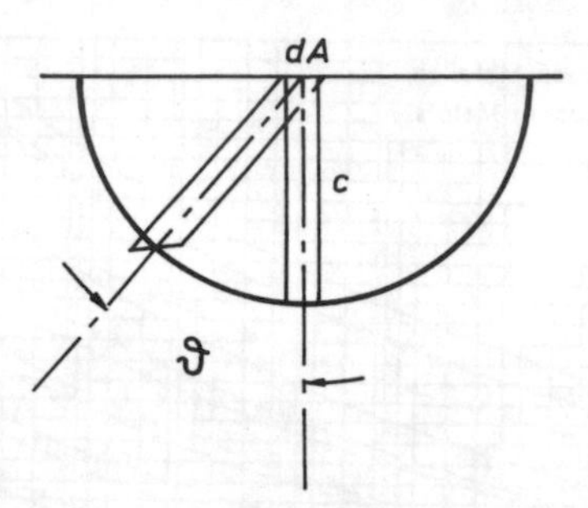


Fig. 17. All atoms with velocity $c$ in the tube of volume $c\cos\vartheta dAn$ can strike $dA$ on the substrate in unit time.

number of evaporated atoms (Eq. (4)). According to Knudsen's law, we obtain

$$\frac{dZ}{d\Omega dAdt} = \frac{P^*}{\sqrt{2\pi mkT}}\cos\theta\ . \tag{7}$$

The thermal velocities of evaporated atoms can be finally obtained. Maxwell velocity distribution yields

$$\frac{dn}{dc} = \frac{4n}{\sqrt{\pi}}\left(\frac{m}{2kT}\right)^{1/3} c^2 e^{-mc^2/2kT} \tag{8}$$

with

$$\langle E\rangle = m\langle c^2\rangle/2 = (3/2)kT \tag{9}$$

and

$$\sqrt{\langle c^2\rangle} : \langle|c|\rangle : \bar{c} = \sqrt{\frac{3}{2}} : \sqrt{\frac{4}{\pi}} : 1 \tag{10}$$

$\bar{c}$ being the velocity for maximum $dn/dc$. In Fig. 18, the velocity distribution is shown schematically for various gases.

The above considerations concern pure systems (single component evaporation). In the following we shall deal with the evaporation of alloys and compounds.

For simplicity, at first alloys are considered as ideal solutions, i.e., the interaction forces between the component atoms A-A, A-B and B-B are the same. Then the vapor pressure of A and B in the solution is given by

$$P'^*_{\mathrm{A}} = c_{\mathrm{A}} P^*_{\mathrm{A}}, \qquad P'^*_{\mathrm{B}} = c_{\mathrm{B}} P^*_{\mathrm{B}} \tag{11}$$

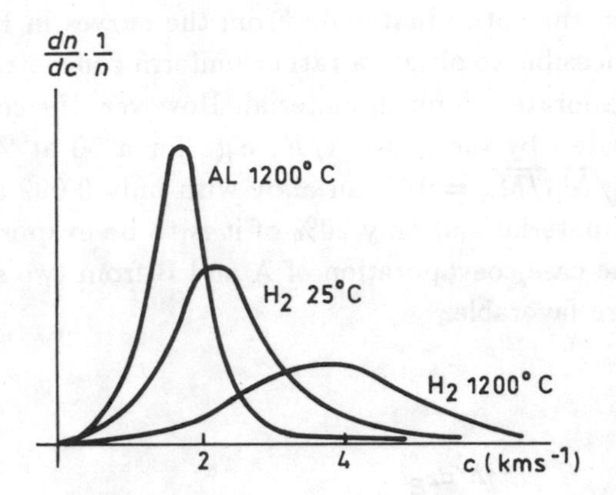


**Fig. 18. Maxwell velocity distribution of atoms. $n$–density of atoms. $c$–velocity.**

where $c_{A,\,B}$ is the molar fraction, subscripts A and B denoting the elements A and B respectively. The ratio of evaporating atoms is then

$$\frac{dZ_A}{dZ_B} = \frac{c_A}{c_B}\frac{P_A^*}{P_B^*}\sqrt{\frac{M_B}{M_A}} \tag{12}$$

Where $M$ is the atomic weight. It is clear that only if $(P_A^*/P_B^*)\sqrt{M_B/M_A} = K = 1$ can we have $dZ_A/dZ_B = c_A/(1-c_A)$, i.e., the vapor shows the same concentration as the source material (congruent evaporation). Otherwise the more volatile component distills rapidly leading to its increasing depletion in the source. We then cannot expect a film with uniform concentration in general. Equation (12) can be considered as a differential equation for the composition of the vapor or of the source. Its solution with $dZ_{A,B} = -dN_{A,B}$ and $N$ being the total number of moles in the source gives

$$\frac{c_A}{(1-c_A)^K} = \frac{c_A^0}{(1-c_A^0)^K}\left(\frac{N}{N_0}\right)^{K-1}, \tag{13}$$

where the subscript and superscript 0 express the starting value of $N$ and $c$ respectively. Equation (13) connects the actual mole fraction of component A in the source with the totally remaining moles. Substituting this into Eq. (12) the composition in the vapor is defined. A plot for an originally 50 at % alloy and different values of $K$, vs the already evaporated fraction of moles $1 - N/N_0$ is shown in Fig. 19. It should be mentioned that immediate mixing by turbulences in the remaining source material has been assumed, otherwise the surface would show a higher depletion of the volatile component than the entire material. From the curves in Fig. 19 it can be seen that it is possible to obtain a rather uniform concentration in the film if we do not evaporate too much material. However, the composition must be precompensated by the factor $1/K$, e.g., for a 50 at % alloy film and $K = (P^*_A/P^*_B)\sqrt{M_B/M_A} = 100$, an alloy with only 0.099 at % A must be used as source material and only 20% of it is to be evaporated. Hence for such an extreme case coevaporation of A and B from two separate sources is certainly more favorable.

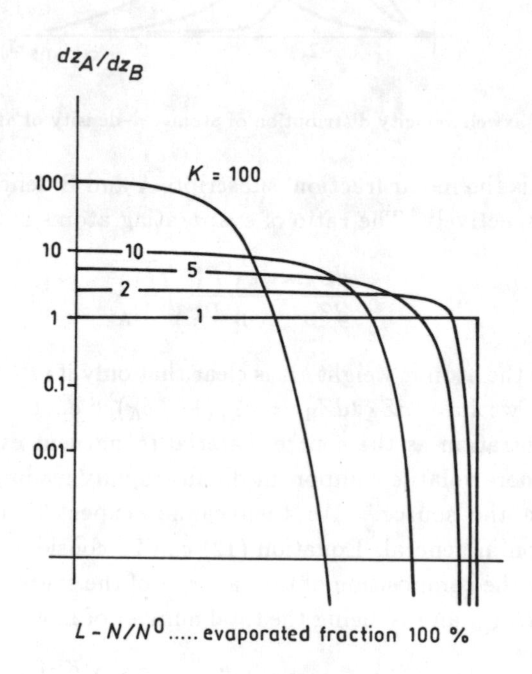


**Fig. 19. The ratio of evaporating atoms of two elements A and B for an originally 50 at % alloy and different $K$, vs the already evaporated fraction of moles $1 - N/N^0$.**

Evaporation of compounds may occur either congruently (e.g., SiO, $MgF_2$) or by dissociation like many II–VI (chalcogenide) semiconductors. In the latter case the reaction of the dissociated components at the substrate must be assured. This can be achieved by a proper substrate temperature, at which the non-reacted fraction is repelled because of its higher vapor pressure.

### 1.2.2.2. *Technical aspects*

The problem of heating the source material can be solved by a vast variety of methods as shown in Fig. 20. The simplest is by heating in a metallic boat by electric current. Boats of refractory metals such as W, Ta, Mo may react with the melted material to give a low melting eutectic (e.g., Fe in Ta). It does not matter if the boat material is dissolved in the melt since the refractory metals evaporate very little as explained in the foregoing section. For dielectrics, the so-called optically tight sources (Drumhellers) are used in order to avoid splitting.

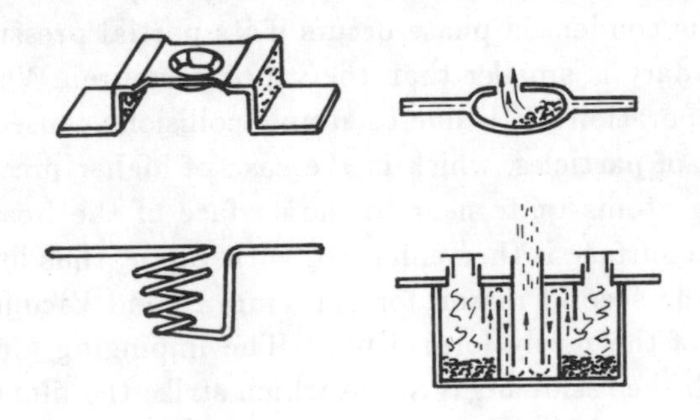

**Fig. 20. The devices for heating the source materials.**

Modern and universally applicable sources are electron guns. Several types are commercially available. The most suitable ones are those with a 270° magnetic electron beam deflection having the advantage of being able to use the whole half space for evaporation, while simultaneously avoiding exposure of the filament and electron optics to the vapor. A reaction of the source material with the crucible is not possible since only a part of it is melted. The temperature obtained by electron bombardment may be as high as required for the evaporation of W, Nb, $Al_2O_3$ etc.

Special methods are sometimes used. Flash evaporation for successive complete evaporation of alloys and compounds is shown in Fig. 21. A properly chosen (small) particle size leads to the consecutive deposition of monolayers, thus enabling the formation of homogeneous films even if the vapor pressure of the components differs considerably. Arc-evaporation for carbon films is a technique which is mostly used for electron microscopic replicas.

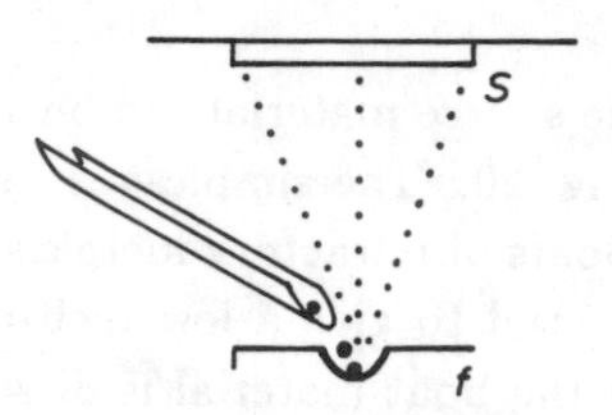


**Fig. 21.** Arrangement for flash evaporation. *S*–substrate. *f*–heating device.

### 1.2.3. *Transport of particles from source to substrate*[3,4]

As discussed in the preceding chapter, during evaporation a net flux of atoms out of the condensed phase occurs if its partial pressure in front of the phase boundary is smaller than the vapor pressure. Why do we need vacuum for evaporation? It is due to atomic collisions caused by the finite mean free path of particles, which in the case of higher pressure will stop the evaporating atoms quite near to the surface of the boat. The space next to the boat surface is then enhanced with vapor, thus limiting further evaporation. The second reason for applying a good vacuum is to avoid contamination of the newly formed film. The impinging film atoms have to compete with the residual gas atoms which strike the film surface to the amount

$$\frac{dZ}{dAdt} = \frac{P}{\sqrt{2\pi mkT}} ,$$

as already pointed out.

In the following section, the idea of mean free path (MFP), the number of collisions etc. for atoms traversing the space between source and substrate will be briefly reviewed.

From the gas equation, the number of gas atoms per unit volume, $n$, at a pressure $P$ is

$$n = P/kT . \tag{14}$$

This means that at atmospheric pressure and room temperature $n = 25 \times 10^{19}$ particles are found in 1 $cm^3$ and at $P = 10^{-8}$ Torr there are still $3.2 \times 10^8$ atoms in 1 $cm^3$. A simple estimation of MFP as the average distance between two collisions is shown in Fig. 22. Through unit length, an atom of diameter $d$ can strike $nd^2\pi$ gas atoms of equal size. The MFP is thus $1/(nd^2\pi)$. The exact value is

$$\mathrm{MFP} = 1/(\sqrt{2}nd^2\pi) . \tag{15}$$

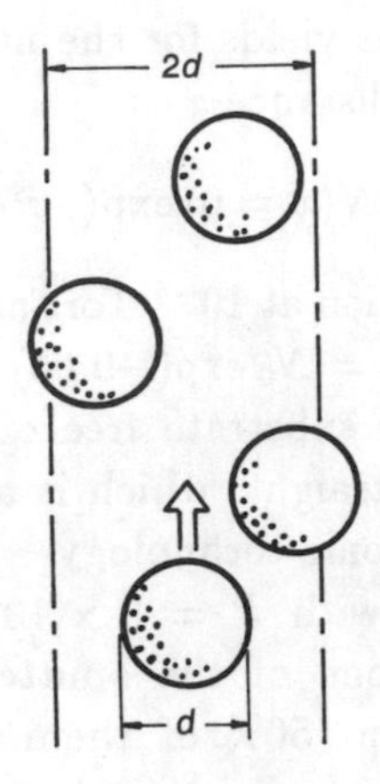


Fig. 22. A simple estimate of mean free path (MFP) for atoms of diameter $d$.

Figure 23 shows the dependence of MFP on pressure for $H_2O$, a large molecule of $d = 4.7 \times 10^{-8}$ cm, and He a small one with $d = 2.2 \times 10^{-8}$ cm. Many other gas atoms show $d$ between these values.

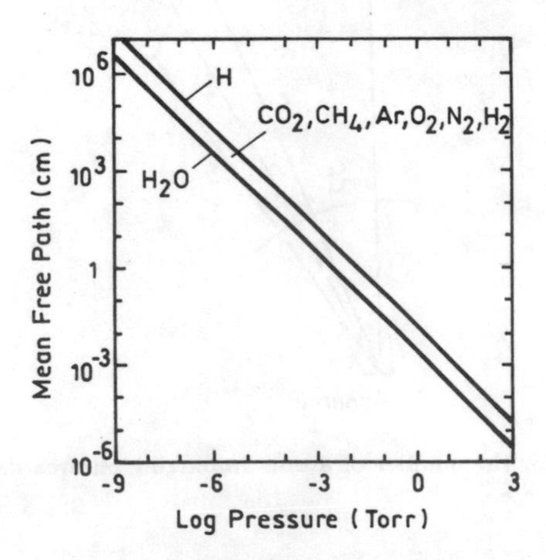


Fig. 23. MFP vs pressure of gases.

A rough estimation of MFP gives the following formula

$$\text{MFP (cm)} = 5 \times 10^{-3}/P \text{ (Torr)} .$$

For example, $P = 10^{-3}$ Torr ... MFP = 5 cm (e.g., low pressure sputtering), $P = 10^{-8}$ Torr ... MFP = 5 km (e.g., vacuum evaporation).

The statistics of collisions yields for the number of migrating particles still free of collision after a distance $x$

$$N(x) = N(x = 0)\exp(-x/\text{MFP}) . \tag{16}$$

This means that for evaporation at $10^{-5}$ Torr and a distance between source and substrate $x = 25$ cm, $N = N_0 \exp(-0.05) \approx N^0$, i.e., almost all of the evaporated atoms reach the substrate free of any collision. Hence their trajectories are absolutely straight, which is a very important feature for masking, e.g., in microelectronic technology.

In sputtering, however, with $P = 5 \times 10^{-2}$ Torr, $x = 10$ cm, $N = N_0 \exp(-100) \approx 0$, i.e., none of the sputtered particles arrive at the substrate without a collision, 50% of them already encounter the first collisions in the first 0.7 mm! Sputtered particles reach the substrate, therefore, from many directions which improves the coverage of steps or relief structure at the substrate.

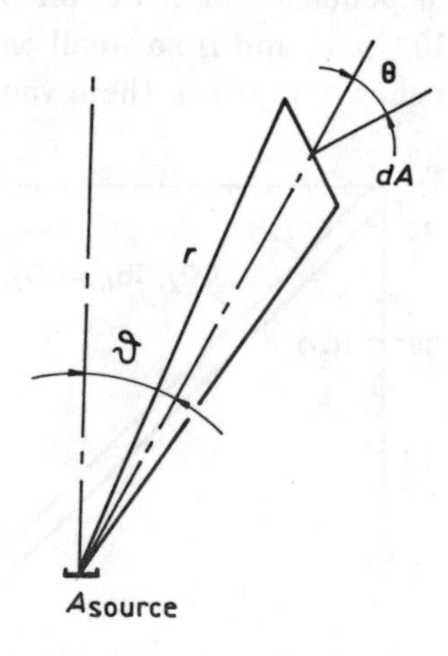


Fig. 24. An estimate for the number of atoms impinging on area $dA_S$ of substrate from area $A_{source}$ of the source.

For evaporation we may now calculate the number of atoms impinging on a substrate exposed to the vapor (Fig. 24)

$$\frac{dZ}{dA_S dt} = \cos\vartheta \cos\Theta \frac{P^* A_{\text{source}}}{r_0^2\sqrt{2\pi mkT}} \tag{17}$$

since

$$d\Omega = dA_S \cos\Theta / r^2 . \tag{18}$$

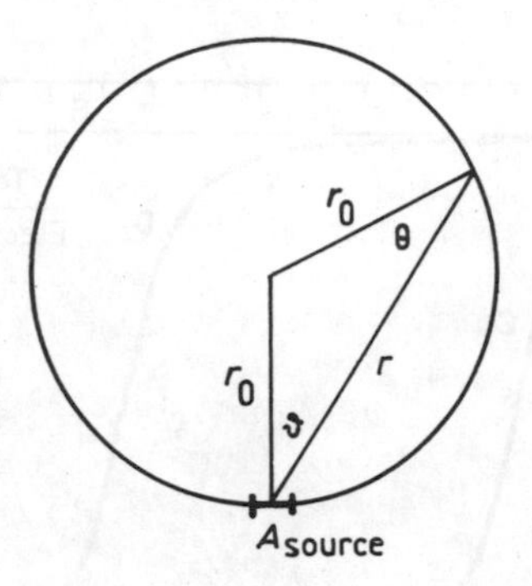


**Fig. 25. Atoms can strike on the sphere surface uniformly.**

Knudsen found that any tangential sphere to the substrate surface is struck uniformly with atoms as can be clearly seen in Fig. 25, where $r = 2r_0 \cos\vartheta$, $\Theta = \vartheta$, and

$$\frac{dZ}{dA_s dt} = \frac{P^* A_{\text{source}}}{4r_0^2\sqrt{2\pi mkT}} .$$

Of course the atoms impinging next to the upper part arrive by grazing incidence. For practical use therefore only the upper part is used for substrate positioning.

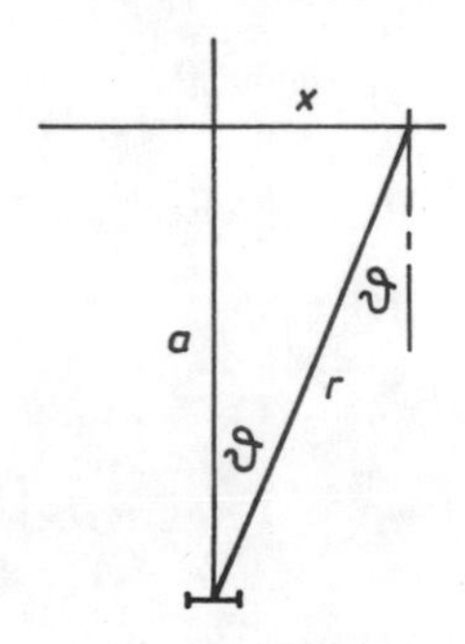


**Fig. 26. Estimate of thickness change along a large flat substrate.**

For practical use the thickness distribution along a large flat substrate is important (Fig. 26) In the case of evaporation Eq. (17) yields

$$Z(x) = Z(x = 0)/[1 + (x/a)^2]^2 . \qquad (19)$$

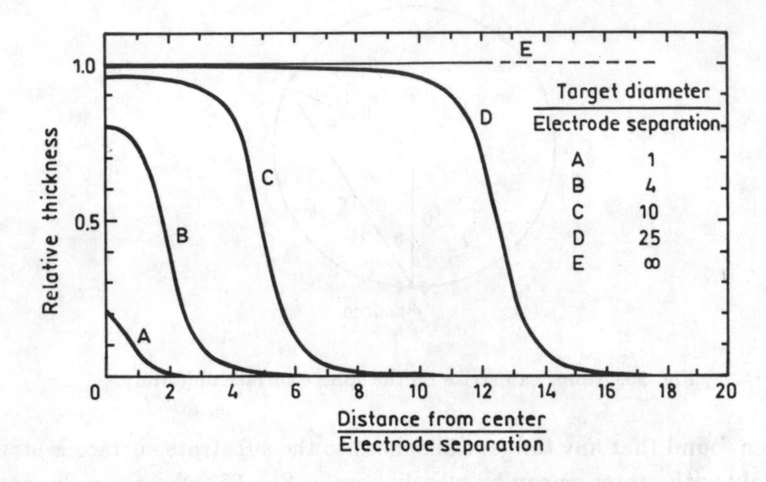


Fig. 27. Thickness distribution along the flat substrate.

For a relative eccentricity $x/a = 0.1$ a deficit in thickness of about 2% will result. For sputtering, the empirical curves in Fig. 27 show the thickness vs distance from the center for various target size.

# CHAPTER 2

# THEORIES OF NUCLEATION AND FILM GROWTH[3,4]

In this chapter the processes of film formation starting from the first atom impinging at the substrate surface to the final film are discussed.

Quantitatively the following steps are involved in film formation:

1) thermal accommodation of the "hot" vapor atoms
2) binding on the substrate surface
3) surface diffusion
4) cluster formation
5) growth of supercritical clusters to islands
6) coalescence of islands
7) growth of the continuous film.

## 2.1. Adsorption

Vapor atoms strike the substrate with an average kinetic energy of $3kT_{\text{source}}/2$. This amount is dissipated by the cold substrate within one or two periods of its lattice vibrations ($10^{14}$ Hz) for most of them. They are trapped in the surface potential: adsorption. The probability of obtaining sufficient thermal energy to surmount the adsorption potential is $\exp(-E_{\text{des}}/kT)$ (Fig. 28), hence the fraction of desorbing atoms per unit time is

$$n_{\text{des}}/n_{\text{ads}} = \nu \exp(-E_{\text{des}}/kT) \ , \tag{20}$$

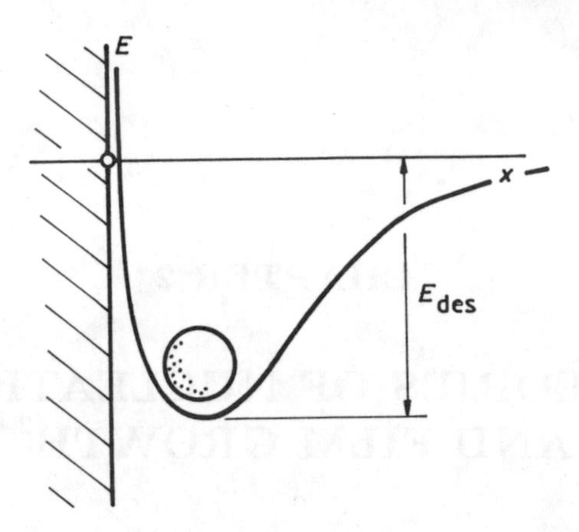


Fig. 28. The potential well for desorption.

$\nu$ is the attempt frequency for escaping, usually taken as the frequency of the vibrating lattice. Equilibrium between impinging and desorbing atoms is achieved when

$$R = \frac{dZ}{dAdt} = n_{\text{des}} \,. \tag{21}$$

The surface population of adatoms is then

$$n_{\text{ads}} = R\frac{\exp(E_{\text{des}}/kT)}{\nu} = R\tau$$

where $\tau$ is mean residence time of an adatom at the substrate surface. For example, $R$ corresponding to a growth rate of 10 Å $s^{-1}$ with an atomic diameter of 1.5 Å gives $R = 5 \times 10^{16}$ atoms/cm$^2$s. For Cu on glass, $E_{\text{des}} = 0.14$ eV, $\exp(-E_{\text{des}}/kT) = 380$, $\nu = 10^{14}$ Hz and $n_{\text{ads}} = 2 \times 10^5$ atoms/cm$^2$. For Ag on NaCl, $E_{\text{des}} = 0.6$ eV, $\exp(-E_{\text{des}}/kT) = 1.1 \times 10^{11}$ and $n_{\text{des}} = 5 \times 10^{13}$ atoms/cm$^2$.

The adatoms may move along the surface, hopping from valley to valley in the surface potential. The frequency of these jumps is

$$\nu_{\text{diff}} = \nu \exp(-E_{\text{diff}}/kT) \,. \tag{22}$$

Denoting the jump distance by $a$, the mean square displacement of an atom hopping statistically during its residence time at the surface is

$$\begin{aligned} \langle l^2 \rangle &= Na^2 = \nu\tau \exp(-E_{\text{diff}}/kT) \cdot a^2 \\ &= \exp[E_{\text{des}} - E_{\text{diff}}/kT] \cdot a^2 \end{aligned} \tag{23}$$

where $N$ is number of jumps.

During its motion on the surface, an adatom may meet others, with which it may react to form a dimer cluster. By reacting with clusters of different sizes, adatoms may be incorporated reversibly if the cluster is of subcritical size.

In the capillarity model of nucleation, clusters are supposed to show bulk properties, i.e., the surface tension and heat of evaporation are those of the macroscopic phase.

The reactions of adatoms with adatoms and with clusters of radius $r$ finally lead to an equilibrium population where clusters of radius $r$ are present with a density given by the law of mass action

$$n(r)/n_{\text{ads}} \doteq \exp[-\Delta G(r)/kT] \tag{24}$$

where $\Delta G(r)$ is the Gibbs free energy of formation of a cluster of radius $r$. This equation holds as long as a critical radius $r^*$ is not exceeded. All clusters $r \geq r^*$ are assumed to incorporate a further atom irreversibly; the reaction is unidirectional since the Gibbs free energy decreases as the reaction proceeds. How do we find this critical size? $\Delta G$ is given by the heat of condensation and the surface energy. In small clusters the surface energy is not completely covered by the condensation energy. The surface energy is given by $\sigma K r^2$ whereas the Gibbs free energy of condensation by $\Delta g_V K' r^3$, where $K$ and $K'$ are constants, $\sigma$ is the coefficient of surface tension and $\Delta g_V$ the heat of condensation per unit volume. Hence $\Delta G = G_{\text{condensed}} - G_{\text{vapor}}$ is positive for rather small clusters where the surface energy is relatively high. The formation of a cluster is associated with an increase of $\Delta G$ up to the critical size. If, however, a cluster has grown to $r^*$ then each further incorporated atom will reduce the Gibbs free energy, i.e., the heat of reaction overcompensates the need for the surface energy increase. The Gibbs free energy of condensation is dependent on the supersaturation of the gas phase:

$$\Delta g_V = -\frac{kT}{V_{\text{atom}}} \ln \frac{P}{P^*}\,, \tag{25}$$

where $V_{\text{atom}}$ is the atomic volume. We may express $P$ and $P^*$ by the numbers $R$ and $R_{\text{ev}}$ of atoms impinging and re-evaporating respectively:

$$\Delta g_V = -\frac{kT}{V_{\text{atom}}} \ln \frac{R}{R_{\text{ev}}}\,. \tag{26}$$

In Fig. 29 $\Delta G$ is depicted for different supersaturations. The critical radius $r^*$ is the value for which $\Delta G_{\max} = \Delta G^*$.

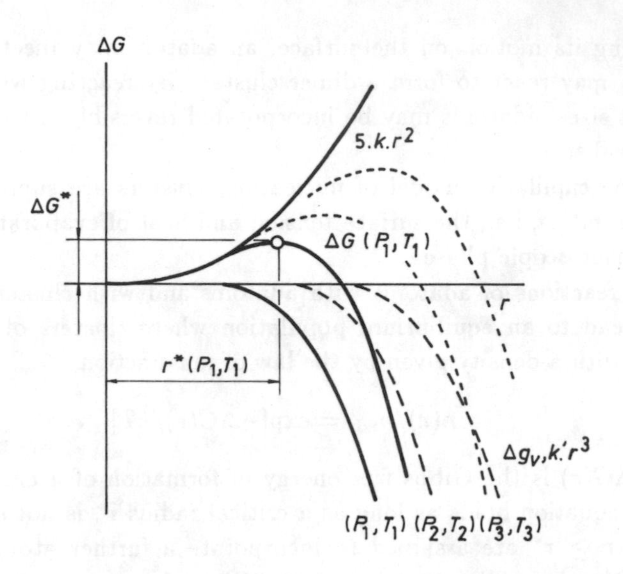


**Fig. 29. Gibbs free energy $\Delta G$ vs cluster radius $r$.**

Furthermore, we call a critical cluster plus one adatom a nucleus. The rate of forming nuclei is given by

$$I^* \propto n(r^*)\nu \exp(-E_{\text{diff}}/kT) \cdot n_{\text{ads}}$$

$$\propto \exp[(2E_{\text{des}} - E_{\text{diff}} - \Delta G^*)/kT] \,. \qquad (27)$$

Arbitrarily one considers a threshold for nucleation when at least one nucleus/cm$^2$s is formed. The associated supersaturation is called critical supersaturation. Experimental results showed that the smallest critical nuclei may contain only a few atoms or at least only one atom. Hence the capillarity model, assuming fairly large aggregates as nuclei, might be overstressed. This difficulty is avoided in the atomistic approach where the Gibbs free energy of the formation of a cluster is replaced by the energy of decomposition $E_i$ of an aggregate into its $i$ single atoms. In this case the law of mass action for the reaction

$$i \text{ adatoms} \overset{E_i}{\longleftrightarrow} \text{cluster } i$$

yields a population with clusters $i$ in a relative density $n_i/N_0$, where $N_0$ denotes number of adsorption sites per area:

$$\frac{n_i}{N_0} = \left(\frac{n_{\text{ads}}}{N_0}\right)^i \exp(-Ei/kT)$$

leading to the rate of formation of clusters containing $i+1$ atoms

$$I_{i+1} \propto n_{\text{ads}} \nu \exp\left(-E_{\text{diff}}/kT\right) \cdot n_i$$

$$\propto \frac{R^{i+1}}{\nu^i N_0^{i-1}} \exp\{[(i+1)E_{\text{des}} - E_{\text{diff}} - E_i]/kT\} \,. \tag{28}$$

This set of equations is consistent only if $I_{i+1} < I_i$. The lowest nucleation rate also defines the critical atomic number. The consideration of embryos and nuclei as an assembly of discrete atoms manifests itself in the discontinuous slope of the nucleation rate as shown in Fig. 30. In some cases this behavior has been proved experimentally.

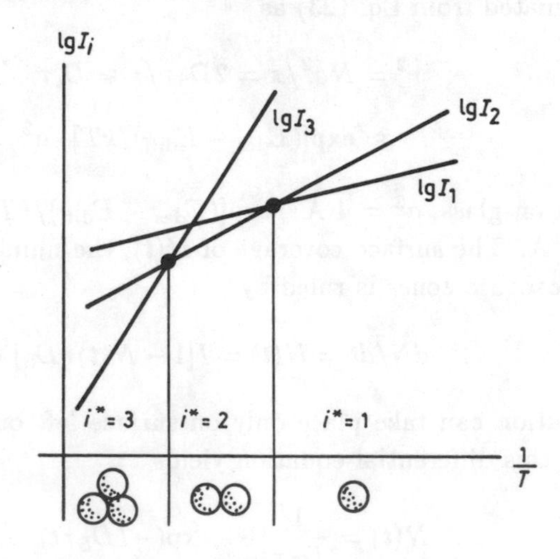


Fig. 30. The atomistic approach of nucleation theory. $I$-rate of cluster formation.

It should be mentioned finally that another atomistic nucleation theory based on decay and formation rates of clusters $i$ was stated by Frenkel and developed by Zinsmeister. It enables the calculation of the time dependence of the population with clusters since it is not based on equilibrium considerations.

## 2.2. Film Growth

For simplicity we shall neglect the decay of super-critical clusters since this process should be rare in comparison to their growth. We shall now estimate the maximum distance in which an adatom can be located, so that it may reach the nucleus during its lifetime. This distance defines the edge of the "capture zone" of the nucleus. It is excluded from the equilibrium consideration before, since it will be gradually depleted of adatoms by irreversible incorporation when they arrive at the nucleus. We have already pointed out that the mean square distance traversed by an adatom during $N$ steps is $Na^2$, which is due to the area successively covered during migration. For isotropy reasons we consider this area as a circle. If there is a cluster on this circle then the adatom will most probably be caught by the cluster. Hence with

$$\langle l^2 \rangle = Na^2 = 2D_S\tau\,, \tag{29}$$

$D_S$ being the surface diffusion coefficient, the radius of the capture zone can be estimated from Eq. (23) as

$$\begin{aligned} r^2 &= Na^2/\pi = 2D_S\tau/\pi \approx D_s\tau \\ &\approx \exp[(E_{des} - E_{diff})/kT] \cdot a^2 \end{aligned} \tag{30}$$

e.g., for Cu on glass, $a^2 = 4\ \text{Å}^2$, $\exp[(E_{des} - E_{diff})/kT] = \exp(0.1/0.0235)$ and $r = 16$ Å. The surface coverage of $N(t)$, the number of nuclei on unit area, with capture zones is ruled by

$$dN/dt = \dot{N}(t) = I[1 - N(t)\tau D_S] \tag{31}$$

since nucleation can take place only on surface left out by capture zones. Solution of this differential equation yields

$$N(t) = \frac{1}{D_S\tau}[1 - \exp(-ID_S\tau t)]\,. \tag{32}$$

This expression holds only for small $t$, as the extension of the nuclei has been ignored so far. Hence we assume that after the time constant $1/(ID_S\tau)$ the surface is completely covered with capture zones. After this time the nucleation period is finished.

The mass increase per unit area $\dot{\mu}(t)$ during nucleation is mainly given by the atoms impinging at capture zones. Taking $N(t)$ from Eq. (32) we obtain

$$\dot{\mu}(t) = N(t)\tau D_S RM/L = (1 - e^{-ID_S\tau t})RM/L \tag{33}$$

where $M$ is the atomic weight, $L$ the Avogadro number and

$$\mu(t) = \dot{\mu}_\infty \left[t - \left(\frac{1 - e^{-ID_S\tau t}}{ID_S\tau}\right)\right] \tag{34}$$

with $\dot{\mu}_\infty = RM/L$. We observe a delayed start of condensation depending on the temperature of the substrate (Fig. 31), corresponding to the time needed for nucleation. For complete condensation, $R$ must be such that equilibrium of adatoms and clusters cannot establish, i.e., (see Eq. (21))

$$R \geq 1/(a^2\tau) \tag{35}$$

We may also roughly estimate when we may expect the deposit not to form a crystalline layer. The formation of crystalline nuclei is inhibited if the overgrowing layer is deposited before atomic jump to an equilibrium position is possible:

$$R \geq (1/a^2)\nu \exp(-E_{\text{diff}}/kT) \tag{36}$$

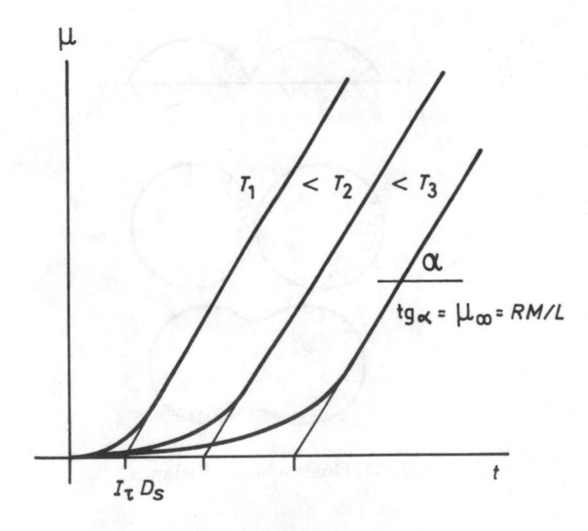


Fig. 31. Mass per unit area $\mu$ during nucleation vs time $t$, showing a delayed start of condensation dependent on the temperature of substrate.

## GROWTH OF NUCLEI AND COALESCENCE OF ISLANDS

After the substrate is completely covered with capture zones, no further nucleation will occur. Islands will grow by incorporation of adatoms diffusing across the substrate or, especially if islands are already large, by direct bombardment with vapor atoms. Particularly interesting is the stage where islands grow together. Let us consider two cap-shaped islands after they touch (Fig. 32). Strong capillarity forces are present next to this zone which is then quickly filled. The islands minimize their surface energy by changing the shape mainly to single cap-shaped aggregates especially when they are still very small. *In situ* electron microscopic observations of this stage show that coalescing islands show a behavior similar to liquid droplets, even if the islands previously revealed crystallographic facets. The facets mostly disappear during coalescence. By reducing the island surface uncovered parts of the substrate reappear where secondary nucleation may take place. After the island has got a proper form the grain boundary separating it with the former ones is able to move (Fig. 33). Parts with energetically less favorable orientations are swallowed. Finally many islands have grown together giving a labyrinthic coverage of the substrate which changes to a continuous film by taking up further atoms.

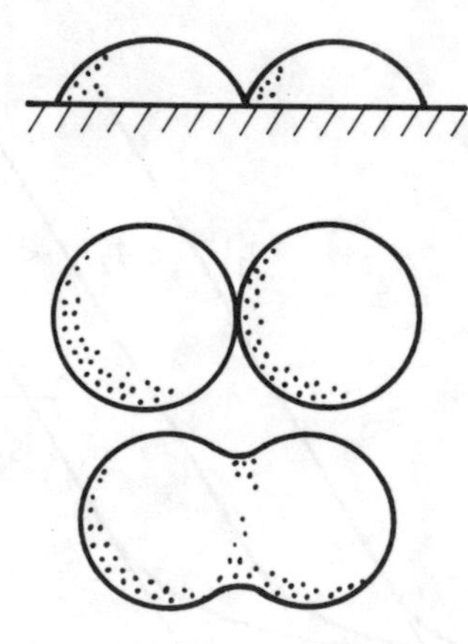

Fig. 32. Coalescence of islands.

## 2.3. Epitaxial Growth[5]

Epitaxy in general is a phenomenon where a relation between the structure of the film and the substrate exists. Commonly, however, epitaxy is understood as the growth of a single crystalline layer on a single crystal surface. In particular, we call the growth of a single crystalline film on a single crystalline substrate of the same material iso-, auto-, or homoepitaxy, which is a rather non-surprising phenomenon. If a single crystalline film grows on a different single crystalline substrate we call the phenomenon heteroepitaxy. This phenomenon, however, is not yet described by a uniform theory and hence not completely understood. Here we gather only

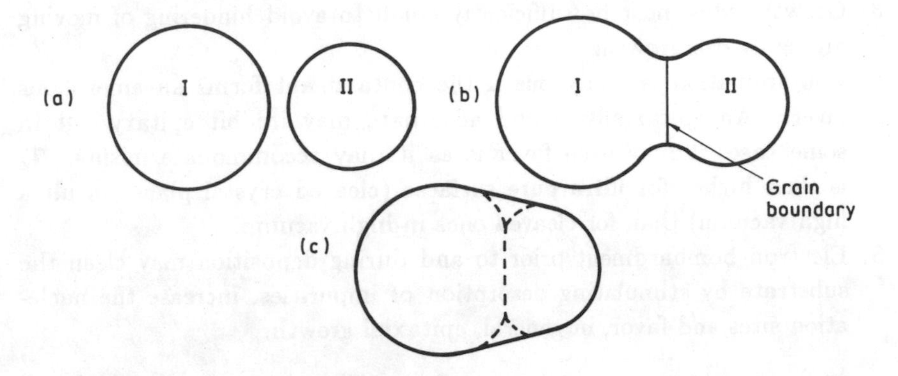


**Fig. 33. When a grain boundary occurs as at (b), it cannot move laterally(c) since its area would increase. Deposition occurs until stage is reached, when the boundary is able to move.**

the empirical findings, which we shall try to make plausible. Two important features are associated with heteroepitaxy:

1) Substrate structure: geometrical compatibility between the substrate surface structure and one lattice plane of the deposit favors epitaxy but is by no means a necessary criterion. Large positive and negative misfits exceeding 10% have been reported to occur in some epitaxial systems. The misfit is accommodated by a proper dislocation network next to the interface or by elastic stress.
2) Substrate temperature: above a more or less well defined (elevated) substrate temperature $T_e$ good epitaxy is obtained. $T_e$ depends on the deposition rate and the surface contamination. The reason for the general need of higher temperature is obviously the reduction of surface contamination by desorption, the enhancement of surface mobility of atoms to reach the favorable sites, and the enhancement of diffusivity in the deposit thus favoring recrystallization and defect annihilation. Nuclei are larger at elevated temperatures. For a defined relation between nucleus and substrate orientation the nucleus must consist of at least 3 atoms for a 3-fold symmetry and 4 atoms for a 4-fold symmetry. Only in such a case will the nuclei show the same orientation and can coalesce without misfits.

3) Growth rates must be sufficiently small to avoid hindering of moving atoms by overgrowth.
4) Contamination is dangerous if the contaminant forms an amorphous layer. An epitaxially bound adsorbate may inhibit epitaxy but in some cases it may even favor it as it may accommodate misfits. $T_e$ is then higher for ultra pure surfaces (cleaved crystal planes in ultra high vacuum) than for cleaved ones in high vacuum.
5) Electron bombardment prior to and during deposition may clean the substrate by stimulating desorption of impurities, increase the nucleation sites and favor, in general, epitaxial growth.

The strongest and universal method of epitaxy is molecular beam epitaxy (MBE). Recently a new technology called CBE (chemical beam epitaxy) which combines MBE with MOCVD (metal organic chemical vapor deposition) has shown very interesting results.[6]

# CHAPTER 3

# CONTROL AND MEASUREMENT OF FILM THICKNESS[3,4]

The goal of any technology is to bring out products of defined and reproducible properties. In thin film technology the desired properties $E_i$ usually depend on many parameters, but most significantly on thickness:

$$E_i = E_i(D, \dot{D}, \ldots, P, P_j, \text{ substrate})$$

where $D$ denotes thickness, $P$ the pressure of the residual gases. Hence methods for *in situ* control and measurement of the film thickness are required in thin film production. Film thickness and/or mass of the deposit may be measured by many techniques which we shall discuss in some detail in this section.

## 3.1. Rate and Thickness Monitors

### 3.1.1. *Measurement of vapor density* (Fig. 34)

The instantaneous value of the density of the evaporating atoms are measured during deposition, thus determining the rate of atoms impinging at the substrate. As the method is not accumulative the film mass per unit area or the thickness must be calculated by integration. In an ionization gauge exposed to the vapor the evaporated atoms are ionized by the thermionically generated electrons when they are accelerated against the anode. The ions move towards the collector electrode where they are discharged. The respective current $I_i$ is proportional to their number, which

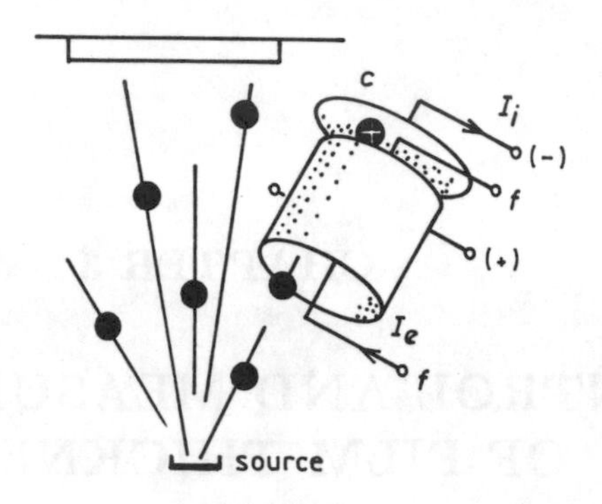


**Fig. 34. An ionization gauge for monitoring film thickness via the measurement of vapor density. $f$–filament. $c$–collector electrode. $I_i$–ion current. $I_e$–ionizing electron current.**

corresponds to the particle density $n$ in the vapor stream and to the ionizing electron current $I_e$

$$I_i \propto I_e n \tag{37}$$

The particle density $n$ in the vapor is related to the growth rate

$$\dot{\mu} = \frac{M}{L}\frac{dZ}{dAdt} \tag{38}$$

by

$$\frac{dZ}{dAdt} = n\langle|c|\rangle \tag{39}$$

which yields, after substituting for the mean particle velocity $\langle|c|\rangle$,

$$\dot{\mu} = \text{ const. } \frac{I_i}{T_e}\left(\frac{8kT_{\text{source}}L}{\pi m}\right)^{1/2} \tag{40}$$

where "const." includes the geometrical factors accounting for the difference in substrate and gauge positions. The disadvantage of this method is the dependence of the result on source temperature and on the residual gas pressure in the vacuum chamber giving a contribution to the ion current. The relative accuracy in the integrated value $D = (1/\rho)\int \dot{\mu} dt$ is about 10%.

### 3.1.2. *Balance methods* (Fig. 35)

These methods are accumulative and use microbalances in different setups. As an example a sensitive current meter is shown, to whose pointer a very light substrate is mounted. The force given by the weight of the condensed

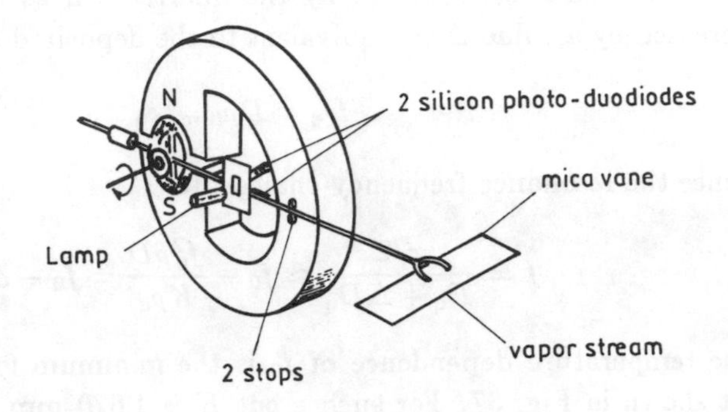


**Fig. 35. Microbalance for monitoring film thickness.**

matter can be compensated by a suitable current through the coil. It can be directly calibrated in terms of the deposited weight.

Micro-balance methods attain a sensitivity of less than 1/100 of a microgram, thus enabling one to measure the mass of less than a monolayer.

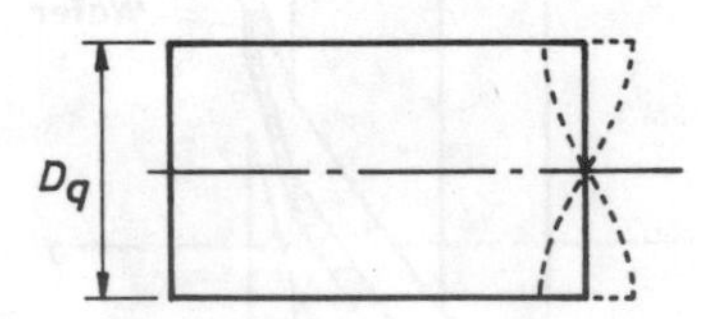


**Fig. 36. Quartz used as a piezo electric resonator operated in a shear mode. $D_q$–its thickness.**

### 3.1.3. *Vibrating quartz method*

This is a method of dynamic weighing by increasing the inertia of a mechanically oscillating system with deposit, thus decreasing its resonance frequency. Quartz is used as a piezo-electric resonator. It is operated in a shear mode with a first order resonance frequency (Fig. 36)

$$f_0 = c/(2D_\mathrm{q}) = K/D_\mathrm{q} \tag{41}$$

where $D_\mathrm{q}$ is the quartz thickness, $c$ the propagation speed of shear wave given by $c = G/\rho_\mathrm{q}$,where $G$ is the shear modulus and $\rho_\mathrm{q}$ the density of quartz. For thin deposit the original antinodes remain unchanged, the additional mass is registered by the quartz such as if its thickness were increased by a value $\Delta D_q$ equivalent to the deposited mass,

$$D_q = D\rho_{\text{film}}/\rho_q \tag{42}$$

Hence the resonance frequency changes to

$$f = \frac{c/2}{D_q + \Delta D_q} \approx f_0 - \frac{f_0^2 \rho D}{K\rho_q} = f_0 - \Delta f \ . \tag{43}$$

The temperature dependence of $f_0$ is the minimum for the so-called A-T cut shown in Fig. 37. For such a cut $K$ is 1.670 mm MHz. Typical data for two thicknesses are listed as follows:

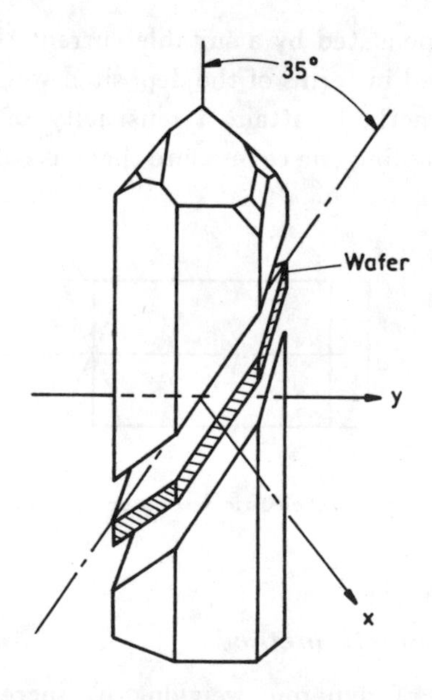


Fig. 37. Quartz crystal with AT cut.

| $D_q$ | $f_0$ | $df/d\mu$ | $df/dD(\rho = 1 \text{ g cm}^{-3})$ |
|---|---|---|---|
| 1.67 mm | 1 MHz | 2.2 Hz/$\mu$g cm$^{-1}$ | 0.022 HzA |
| 0.28 | 6 | 81.5 | 0.815 |

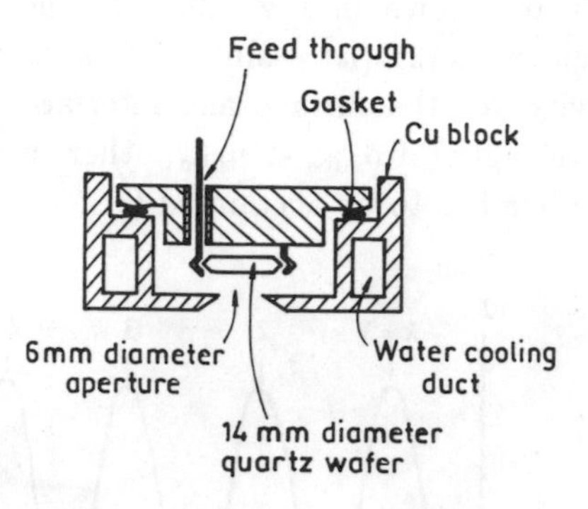


Fig. 38. Practical construction of a quartz thickness monitor.

Practical construction and block chart of circuit are shown in Fig. 38 and 39 respectively. The sensitivity is mainly given by the mechanical limit of the quartz thickness and the stability of the reference oscillator: usually $10^{-7}$ g cm$^{-2}$ corresponding to $D = 1$ Å ($\rho = 10$).

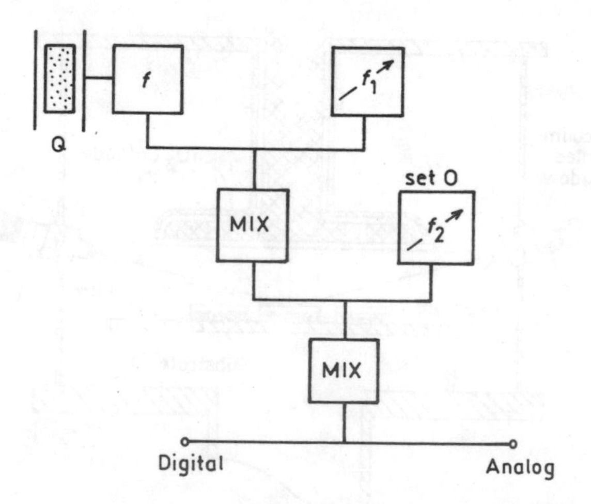


Fig. 39. Block chart of circuit for a quartz thickness monitor.

### 3.1.4. *Optical monitoring*[7]

This method is based on the generation of light interference by the growing film and is hence useful for transparent films only. An example used inside a sputter system is shown in Fig. 41. The changes in intensity during film growth are shown in Fig. 40. For the interference maxima the difference in light path, $2D(n^2 - \sin^2\theta)^{1/2}$, $D$ being the film thickness and $n$ the refractivity, for the surface and interface reflection is $k\lambda$ ($k = 1, 2, 3, \ldots$, $\lambda =$ wavelength) if $n_{\mathrm{subs}} < n_{\mathrm{film}}$, otherwise the phase change at film/substrate interface has to be accounted for.

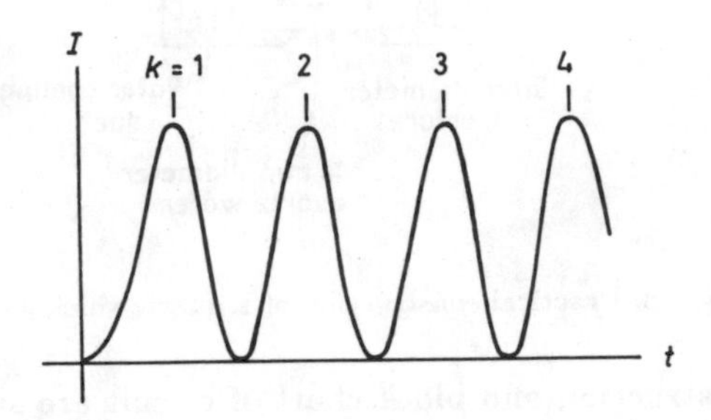


**Fig. 40.** The change of intensity $I$ for a optical thickness monitor. $t$–momentary thickness during film growth.

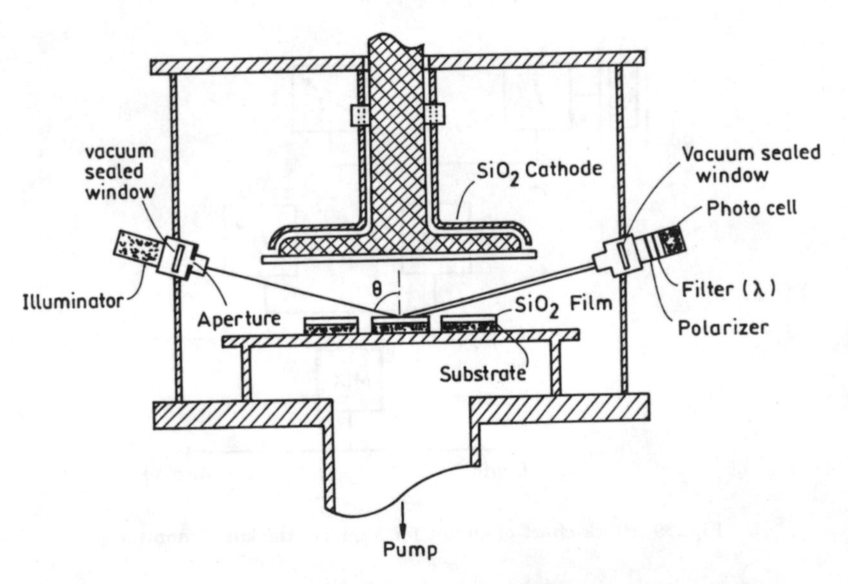


**Fig. 41.** An example of optical thickness monitor inside a sputter system.

### 3.1.5. *Other monitors*

In principle any thickness-dependent film property can be used for monitoring thickness. However, as many properties depend sensitively also on other parameters as, e.g., pressure, deposition rate, temperature etc., they can be successfully correlated with thickness only if the latter are kept constant.

Resistivity and capacity are often used for monitoring thicknesses.

## 3.2. Measurement of Film Thickness

Many monitors must be calibrated. This is usually done by comparing the monitor signal with the result of an independently performed measurement of thickness of the finished film. Absolute thickness measurement is possible by employing optical interference.

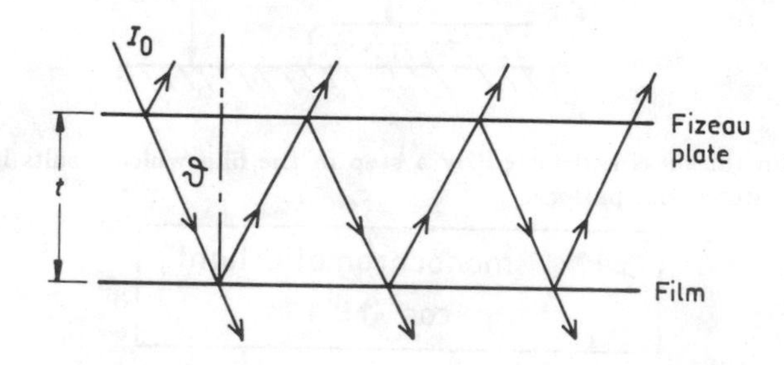


Fig. 42. Optical interferences for the absolute thickness measurement.

### 3.2.1. *Optical film thickness determination (Interferometry)*[7]

This method can be carried out by using a "Fizeau" plate, which enables multiple reflection leading to sharp interference (Fig. 42). The resultant intensity is

$$I_{\rm R}(\Delta\varphi) = I_0 \left\{ 1 - \frac{T^2}{1-R^2} \left[ \frac{1}{1 - F\sin^2(\Delta\varphi/2)} \right] \right\} \tag{44}$$

where $I_0$ denotes the intensity of the primary beam, $T$ transmittance, $R$ the reflectivity, $F = 4R/(1-R)^2$ and

$$\Delta\varphi = (2\pi/\lambda)2t\cos\vartheta\ . \tag{45}$$

Equation (44) rules the dependence of the total reflected intensity on the distance between the film surface and the Fizeau plate. Film thickness is detected by a step in the film which results in a shift of the minima of the interference pattern (Fig. 43). Two methods are in use: the first is TOLANSKY interferometer (Fig. 44) and the second is FECO (Fringes of equal chromatic order) (Fig. 45). In the first method the relative fringe shift $x/l$ is related to $D$ by $D = \lambda x/(2l)$. The resolution in $D$ is 10–30 Å. As for the second method, the fringe shift is $\Delta\lambda_N = 2DN$, where $N$ is the order of interference. It can be evaluated from the distance of the fringes. The resolution is 1 Å.

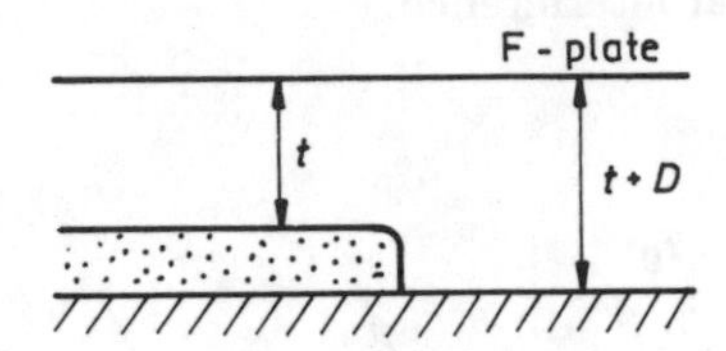


**Fig. 43.** Film thickness is detected by a step in the film which results in a shift of the minima on interference pattern.

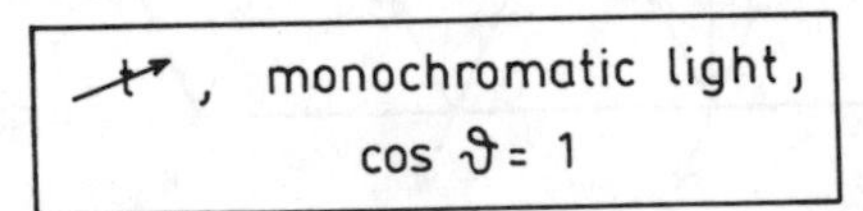


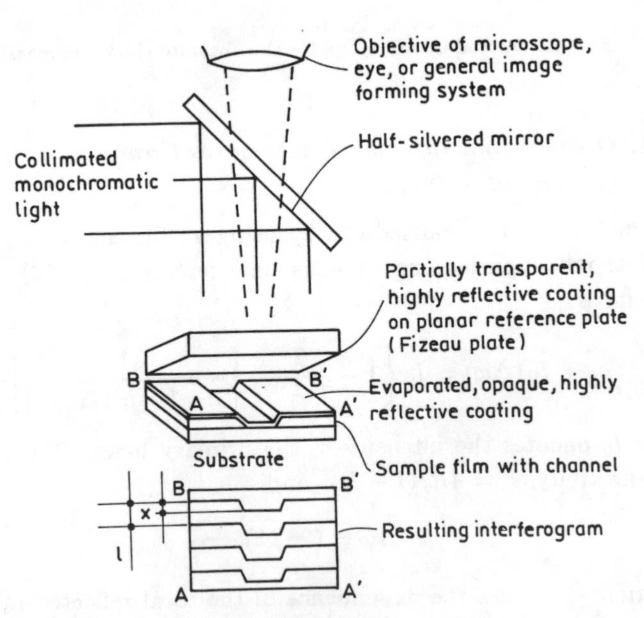


Fig. 44. A TOLANSKY interferometer.

For achieving the same reflectivity on the film and the step bottom both have to be coated with an overlayer (mostly silver).

Other interferometers can be used in the same way showing less accuracy since they mostly use two-beam interferometry, thus giving diffuse fringes. The smallest detectable thickness is usually in the range of 100–300 Å.

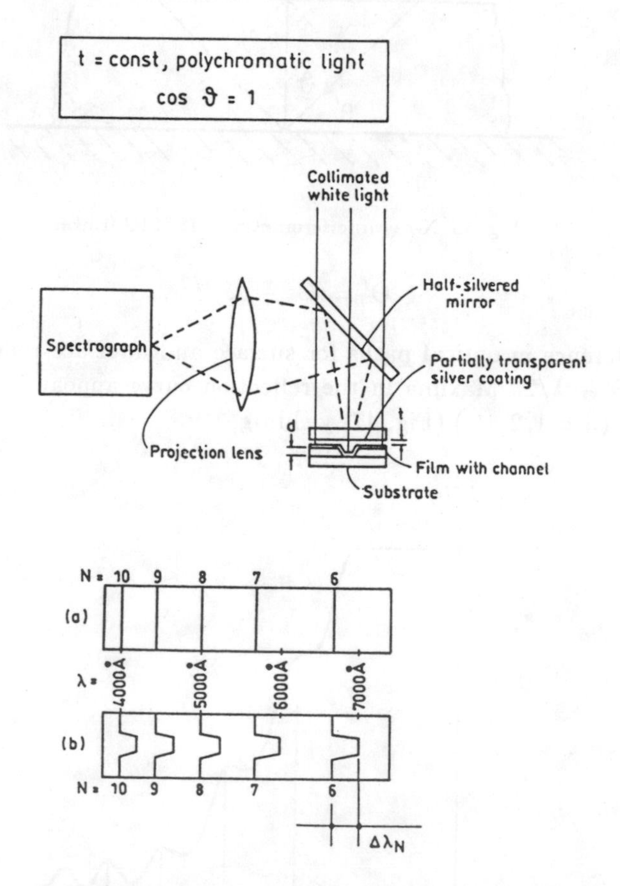


Fig. 45. A FECO (Fringes of equal chromatic order) interferometer.

### 3.2.2. *X-ray interferometry* (*Kiessig fringes*) (Fig. 46)[8]

On grazing incidence X-rays are reflected from and refracted through flat surface. The index of refraction is only slightly different from unity: $n = 1-\delta$, $\delta \approx 10^{-4}$ (e.g. for Ag, $\delta = 31 \times 10^{-6}$). Snell's law gives $\sin\vartheta/\sin\vartheta' = 1-\delta$ and, after some transformation,

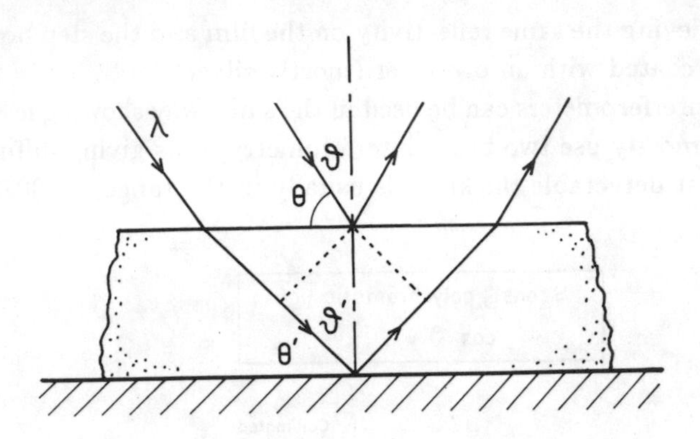


Fig. 46. X-ray interferometry (KIESSIG fringes).

$$\Theta' = (\Theta^2 - 2\delta)^{1/2}\,. \tag{46}$$

The difference in optical paths for surface and interface reflection is $\Delta = 2D\sin\Theta' + \lambda/2$. Maxima in the reflection curve appear at angles where $\Delta = n\lambda$ $(n = 1, 2, \dots)$ (Fig. 47) yielding

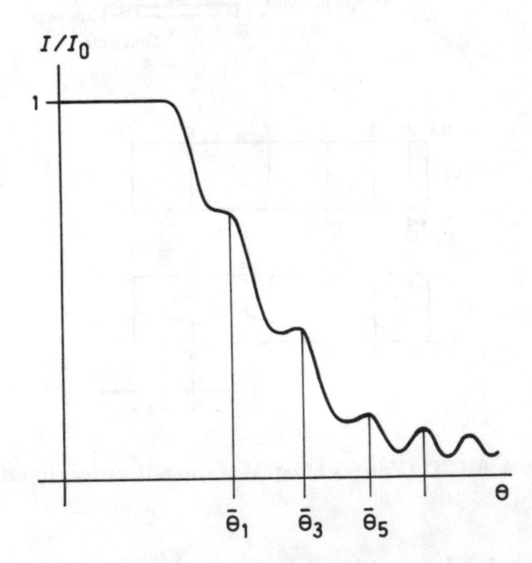


Fig. 47. Reflection curve for X-ray interference.

$$D = K\frac{\lambda}{4}\frac{1}{\sqrt{\overline{\Theta}_K^2 - 2\delta}}, \qquad (K = 1, 3, 5\dots) \tag{47}$$

The correct $K$ can be found by trial and error. The scattering in $D$ should be minimum for the correct $K$. The method is very useful for films below 1000 Å. The resolution is 1–5 Å.

### 3.2.3. *Stylus methods* (Fig. 48)

A diamond tip is moved along the film surface while its vertical displacement is electronically enlarged by a factor of $10^6$ and recorded. From the film edge the thickness is found directly as the step height detected by the stylus. The resolution is 10–20 Å.

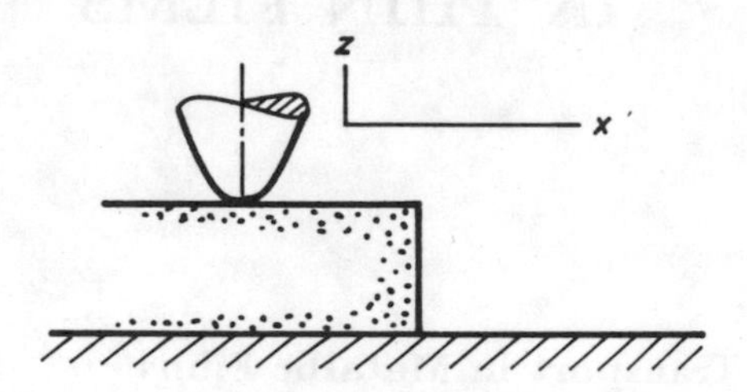


Fig. 48. Stylus method for thickness measurement.

# CHAPTER 4

# ELECTRICAL CONDUCTION IN THIN FILMS[9]

## 4.1. Electron Transport in Metallic Films

We have seen that films are usually discontinuous in an early growth stage. Metal films hence do not provide a continuous bridge for electric current in the island stage. The conduction mechanism in this stage are of course different from metallic conduction but mainly identical to the mechanism characteristic of insulating films sandwiched by metal films. In the following sections the basic principles of the mechanisms in each growth stage are briefly outlined, followed by a section dealing with the application of metal films as conductors and resistors.

### 4.1.1. *Conduction in discontinuous films* (Fig. 49)

In metal films the resistance along the film plane is the important property. For an island film a conductivity, averaged along this plane, is assigned to its mean geometrical thickness $D = \mu/\rho$, $\mu$ being the mass per unit area and $\rho$ the density:

$$\sigma = l/(RbD)$$

where $R$ is the resistance, $b$ the breadth, $l$ the length, $D$ the average thickness, $r$ the average island radius and $d$ the average island spacing. The experimental findings are:

1) $\sigma$ smaller than that for a continuous film,
2) $\sigma \propto \exp(-A/kT)$,
3) $\sigma = \sigma(F)$ for large $F$, $F$ being the field strength,
4) $\sigma = \sigma(r, d)$.

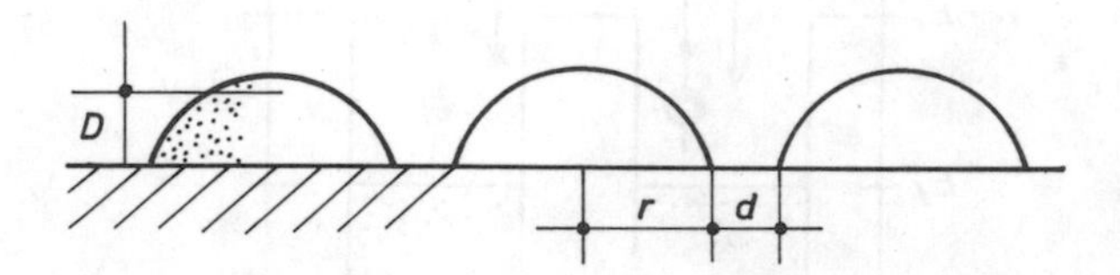


**Fig. 49. Electric conduction in discontinuous film.**

They can be explained by two mechanisms, in principle simultaneously present but each dominating at different temperatures $T$ and geometrical situations $r, d$.

1) *Thermionic emission of electrons either into vacuum or into the conduction band of the insulating substrate*

An applied field drives electrons from island to island thus generating a current. In Fig. 50 the situation without and with an electric field is shown for emission in vacuum. The potential to be surmounted is the work function $W$. It may be reduced by the image force generated in the island from which just left the electron (Schottky effect) and, if island spacing is small, by the superposition of the image forces from the neighboring islands. Richardson equation defines the saturation current density of emitted electrons and hence the conductivity.

$$\sigma \propto dT \exp\left\{-\frac{1}{kT}\left[W - \sqrt{e^3 F} - \text{const.}\ \frac{e^2}{d}\right]\right\} \tag{49}$$

where $\sqrt{e^3 F}$ is the Schottky term and const. $e^2/d$ the image force superposition. This mechanism dominates for $d \geq 100$ Å and $T \geq 300$ K. As for the thermionic current via the substrate's conduction band, it is referred to the section: electron transport through insulating films.

2) *Tunneling via vacuum and via substrate* (Fig. 51)

As a quantum effect it requires the solution of Schrödinger's equation for the boundary conditions given by the island structure. Inside the island the

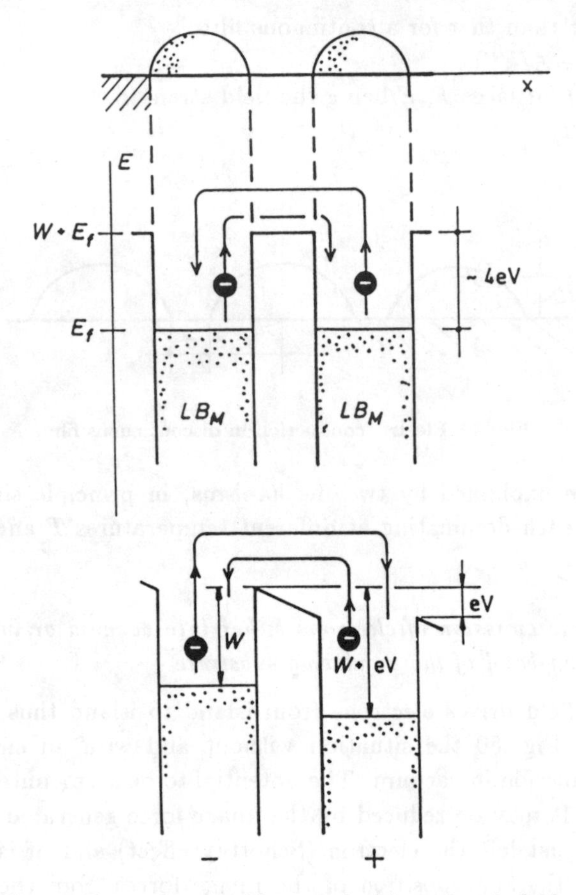


Fig. 50. Energy bands without and with an electric field.

wave function is approximated by plane waves, outside by damping ones. The continuity of its amplitude at the boundary leads to a transmittance $\tau$ mainly ruled by an exponential,

$$\tau \propto \exp\left[-\frac{4\pi d}{h}\sqrt{2m(W-E)}\right] \tag{50}$$

$m$ being the mass of an electron. It must be weighed by the distribution of electrons versus their energy states and the number of free states at the particular energy since tunneling conserves energy. For the positive flux this weighing function is $f(E)[1-f(E+eV)]$, while for the negative, $f(E+eV)[1-f(E)]$. We finally obtain after the linearization of the Fermi distribution $f(E+eV)$ for small potentials eV

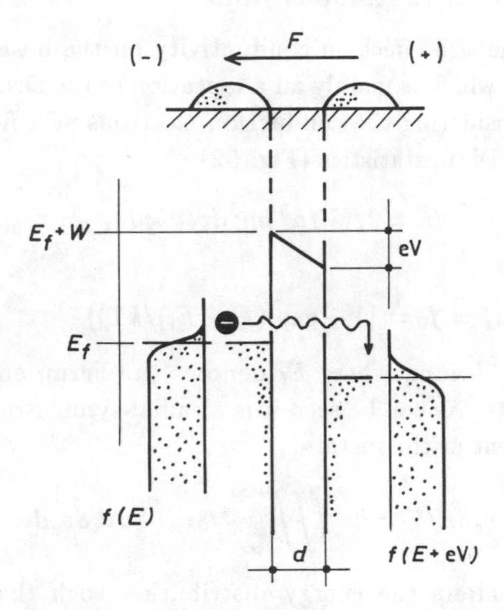


**Fig. 51. Electron tunneling via vacuum and via substrate.**

$$\sigma \propto \frac{d^2}{r} \tau \exp\left(-\frac{e^2/r}{kT}\right) \tag{51}$$

This effect is more dominant for small $d$ as it appears in the exponential. If the islands consist of a few atoms only, then the electron states appear in its quantization which has to be accounted for.

For tunneling via substrate refer to the section: Conduction in Insulators.

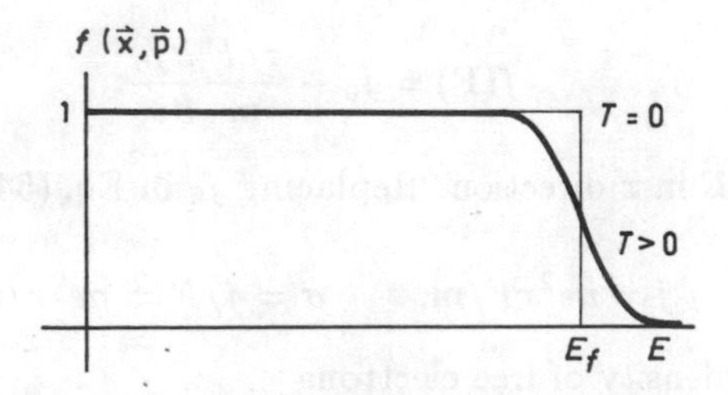


**Fig. 52. Energy distribution of electrons.**

### 4.1.2. *Conduction in continuous films*

Here we discuss the size effect on conductivity on the basis of the Fuchs-Sondheimer theory which is mainly an adaptation of the Drude theory. The latter starts by considering the conduction electrons as a free-electron gas obeying the Fermi Dirac statistics (Fig. 52)

$$dn = 2f(V/h^3)m^3 dv_x dv_y dv_z \tag{52}$$

with

$$f = f_0 = \{1 + \exp[(E - E_f)/kT]\}^{-1} \tag{53}$$

in the case of equilibrium, where $E_f$ denotes the Fermi energy, $v_{x,y,z}$ the velocity components. As can be seen it is a radial-symmetric function, thus giving no net current in the metal

$$j = e \int_N v_x dn/V = 2e \iiint_{-\infty}^{+\infty} f_0 v_x \frac{m}{h^3} dv_x dv_y dv_z = 0 \,. \tag{54}$$

An external field alters the energy distribution such that its spherical symmetry is lost. The influence of external force $\mathbf{F}$ is found by the Boltzmann equation

$$\frac{df}{dt} = \mathbf{v} \cdot \nabla_r f + \frac{\mathbf{F}e}{m} \cdot \nabla_v f + \frac{\partial f}{\partial t} - \left(\frac{\partial f}{\partial t}\right)_{\text{coll}} = 0 \tag{55}$$

for steady state conditions. The collision term is found by a relaxation predicting an exponential relaxation by collision.

$$df(t) = -f(t)dt/\tau \tag{56}$$

$$(\partial f/\partial t)_{\text{coll}} = -(1/\tau)f(t) \tag{57}$$

where $\tau$ is the relaxation time. Inserting this into Eq. (55) we may solve this differential equation for $f$ by introducing several approximations, yielding

$$f(\mathbf{F}) = f_0 - \frac{\tau e F}{m}\frac{\partial f_0}{\partial v_x}$$

if one assumes $\mathbf{F}$ in $x$ direction. Replacing $f_0$ in Eq. (54) by this expression yields

$$j = ne^2\tau F/m, \qquad \sigma = j/F = ne^2\tau/m \tag{58}$$

where $n$ is the density of free electrons

$$n = (8\pi/3)(v_f m/h)^3$$

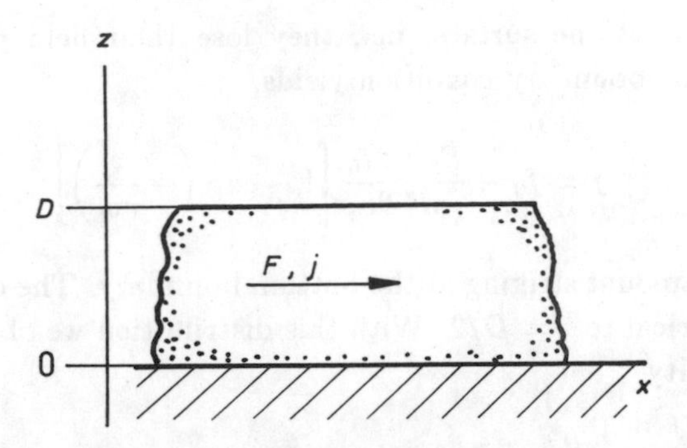


Fig. 53. Current along the film.

where $v_f$ is velocity at Fermi level. $\tau$ is associated to the mean free path by $v_f\tau = \text{MFP}$.

For thin films we have to account for the modification of $f$ by surface collisions of electrons. This is done by setting (Fig. 53)

$$f = f_0 + A(v, z) \tag{59}$$

By neglection of $\partial A/\partial v_x$ after having inserted $f$ into the Boltzmann equation we obtain its solution

$$f(F, z) = f_0 - \frac{Fe\tau}{m}\frac{\partial f_0}{\partial v_x}\left\{1 - K(v)\exp\left[-\frac{z}{\tau v_z}\right]\right\} \tag{60}$$

where $K(v)$ is an integration constant which has to be found from boundary conditions.

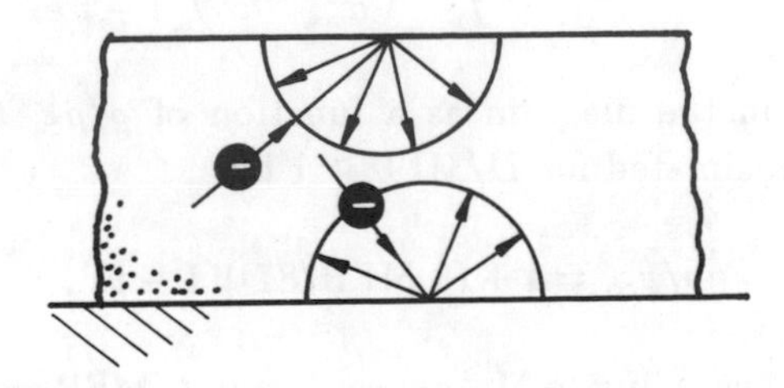

Fig. 54. Diffuse scattering of electrons at the film surfaces.

If we assume diffuse scattering at the surfaces (Fig. 54) then we have to postulate that all electrons start with their equilibrium distribution after their collision at the surface, i.e., they lose their field drift completely. Fitting to this boundary condition yields

$$f = f_0 - \frac{Fe\tau}{m}\frac{\partial f_0}{\partial v_x}\left[1 - \exp\left(-\frac{z}{\tau v_f}\right)\right] , \tag{61}$$

for the half amount striking at the bottom boundary. The other part shows $f$ is symmetrical to $z = D/2$. With this distribution we obtain the average current density

$$\begin{aligned} j &= (1/D)\int_0^D j(z)dz \\ &= j_\infty\left[1 - \frac{3\ \mathrm{MFP}}{8D} + \frac{3\ \mathrm{MFP}}{2D}\int_1^\infty\left(\frac{1}{a^3} - \frac{1}{a^5}\right)e^{-Da/\mathrm{MFP}}da\right] \end{aligned} \tag{62}$$

where $j_\infty$ is the current density for the bulk, a results from integration over $z$. This integral cannot be solved analytically.

$$j/j_\infty = \sigma/\sigma_\infty = 1 - 3\ \mathrm{MFP}/8D \tag{63}$$

is an approximation for $D/\ \mathrm{MFP} \geq 1$, usually valid for films with $D \geq 100$–400 Å, which is the range for the MFP of most metals at room temperature. It is plausible that elastic collision of electrons at the surface (Fig. 55) will not alter the bulk conductivity since the drift is the same before and after scattering. Let $p =$ fraction of elastically scattered electrons, then we obtain

$$\sigma/\sigma_\infty = 1 - (3\ \mathrm{MFP}/2D)(1-p)\int_1^\infty\left(\frac{1}{a^3} - \frac{1}{a^5}\right)\frac{1 - \exp(-Da/\mathrm{MFP})}{1 - p\exp(-Da/\mathrm{MFP})}da , \tag{64}$$

which is shown in the diagram as a function of $\rho/\rho_\infty$ (Fig. 56). Equation (64) is approximated for $D/\mathrm{MFP} \geq 1$ by

$$\rho/\rho_\infty = 1 + (3\ \mathrm{MFP}/8D)(1-p) . \tag{65}$$

This means that according to Mathiessen's rule $1/\mathrm{MFP} = \Sigma 1/\mathrm{MFP}_i$, where $\mathrm{MFP}_i$ are the single effect limitations of the MFP, the size effect on MFP is

$$\mathrm{MFP}_{\mathrm{TF}} = 8D/[3(1-p)] \tag{66}$$

MFP of metals is quite sensitive to temperature as seen below:

| Metal | MFP at −200°C | MFP at 0°C | MFP at 100°C |
|---|---|---|---|
| Li | 955 | 113 | 79 |
| Na | 1870 | 335 | 233 |
| Cu | 2425 | 421 | 294 |
| Ag | 2965 | 575 | 405 |
| Au | 1530 | 406 | 290 |
| Fe | 2785 | 220 | 156 |
| Co | | 130 | 79 |
| Ni | | 133 | 80 |

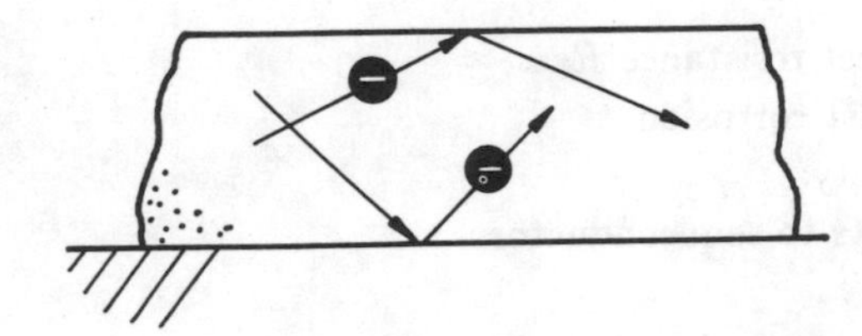

Fig. 55. Elastic collisions of electrons at the film surfaces.

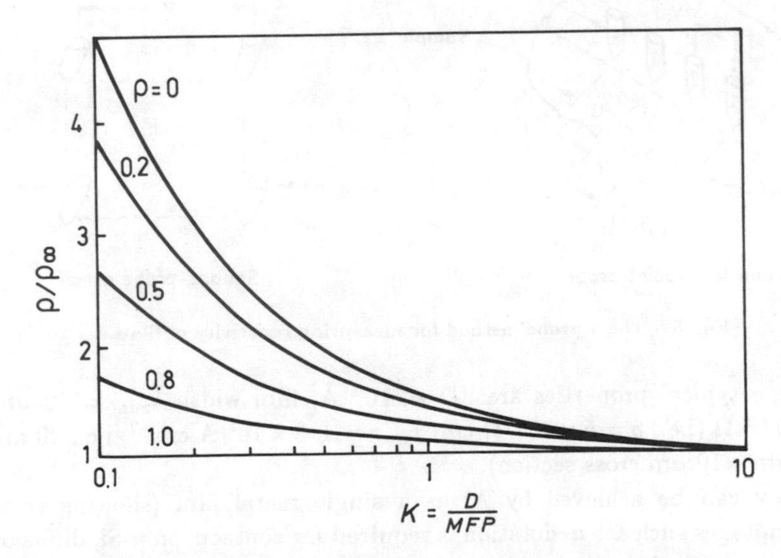


Fig. 56. The effect of film thickness $D$ on electric resistivity $\rho$. $p$–fraction of elastically scattered electrons.

Besides the size effect there is a very significant influence of structural defects on the film resistance. Most effective are point defects mainly quenched in vacancies giving rise to about 1 – 6 Ω cm per at% vacancies. As defects annihilate with time the resistance will change. For practical purposes a pre-aging to stabilize its value is usually required.

### 4.1.3. *Practical aspects and applications*

For practical comparisons of thin films as resistors the quantity $\rho/D = R_s =$ sheet resistance is useful. It is the resistance of any square of the film. It can be measured directly by 4-probe methods on a large area of the film (Fig. 57). Thin films for contacts and interconnections should generally fulfill the following requirements:

1) very small sheet resistance $R_s$.
2) resistive against corrosion.
3) thermal stability.
4) Ohmic contacts to semiconductors.
5) adhesion.
6) good bonding properties.

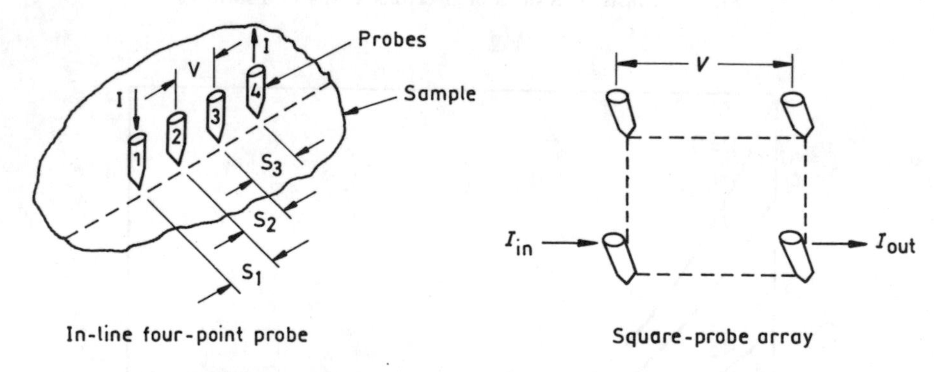


Fig. 57. The 4 probe method for measuring resistivity of films.

Some typical properties are: $D \approx 10^4$ Å, film width $b_{\min} = 10$ μm, $R_s = 0.05$ Ω (i.e., $\rho = 5 \times 10^{-6}$Ω cm, $j_{\max} = 1\text{–}3 \times 10^5$ A cm$^{-2}$, i.e., 20 mA for 1 μm ×10 μm cross section).

They can be achieved by Al as a single metal film (showing some disadvantages such as: $n$–dotation is required for contacts on $n$-Si, diffusion with Au wires). More expensive metallizations are Mo + Au, PtSi + Ti + Pt + Au, etc.

Metal films for resistors should show

1) very high resistivity, and

2) very small TCR (temperature coefficient of resistivity), which are achieved by sputtered Ta doped with a high content of $N$ (thus giving $\rho = 10^2$–$10^4$ $\mu\Omega$ cm, TCR = 0.01% $K^{-1}$), metallic alloys such as NiCr 80/20 flash evaporated ($\rho = 100$ $\mu\Omega$ cm, TCR = 0.01% $K^{-1}$), CrTi 35/65 and cermets which are metal ceramic compounds produced by co-evaporation or co-sputtering.

Special applications of metal and cermet films, especially discontinuous metal films, can be found for gas detectors, thermisters and strain gauges.

## 4.2. Electron Transport Through Insulating Films[10]

Contrary to metal films, conduction across the film is of interest. High field strength is expected, e.g., at a voltage around 1 V and thickness 1000 Å, field strength can reach 0.1 MV $cm^{-1}$.

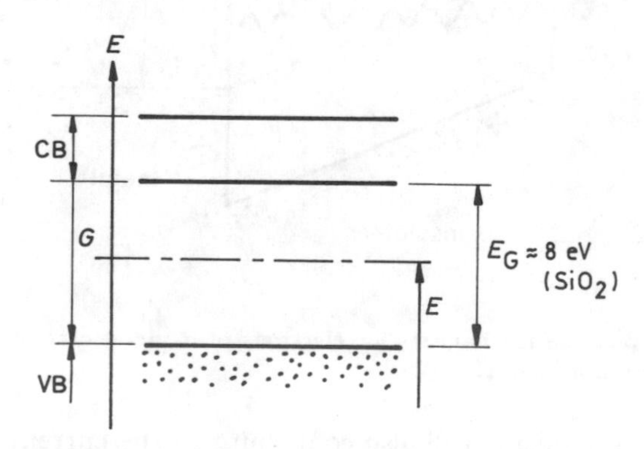


Fig. 58. Simplified band scheme for electron transport through insulating films.

Conductivity of insulators can be measured only when they are enclosed between two metal electrodes. Hence the conduction mechanism is likely to be influenced by such a setup. In any case the band structure plays an important role. The simplified band scheme (Fig. 58) shows a completely filled valence band and a free conduction band at $T = 0$. At room temperature the thermionically created hole-electron pairs are very few since the band gap is so much higher than the thermal energy. Their contribution to conduction (intrinsic conduction) would give resistivity of about $10^{24}$ $\Omega$ cm which is far beyond the measured value. Hence we have to look for other explanations. Figure 59 shows all possible mechanisms for electrons of a metal cathode to reach the anode for a simplified band scheme of the MIM (metal-insulator-metal) structure:

1) Thermionic emission of metal electrons into the conduction band.
2) Tunneling into the conduction band.
3) Tunneling via allowed states in the gap (hopping via traps).
4) Tunneling through the gap.

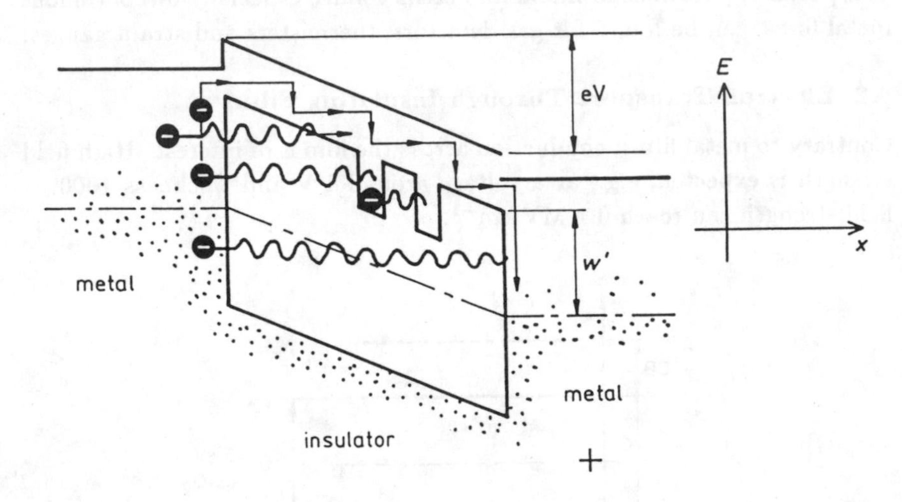


**Fig. 59. All possible mechanisms for electrons of a metal cathode to reach the metal anode through a insulator.**

These mechanisms will also contribute to the current in island films as mentioned before.

As it is in general easily possible to make an insulator sufficiently thick to assure good isolation properties no remarks concerning their applications is made as long as only insulation is required. A physically interesting application, although up to now not in a practical state, with regard to conduction through an insulator film is the TUNNEL EMITTER (Fig. 60) which is just a simple MIM structure. For voltage bias $eV > \Psi$ the electrons first tunnel into the conduction band of the insulator and further into the anode as hot electrons which can escape if they do not lose energy by collisions. Hence the anode has to be less thick than the MFP of the electrons. The tunnel junction then constitutes a COLD CATHODE

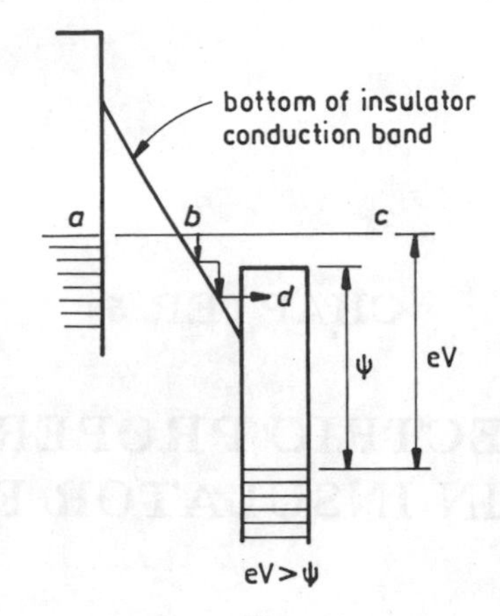


Fig. 60. A tunnel emitter.

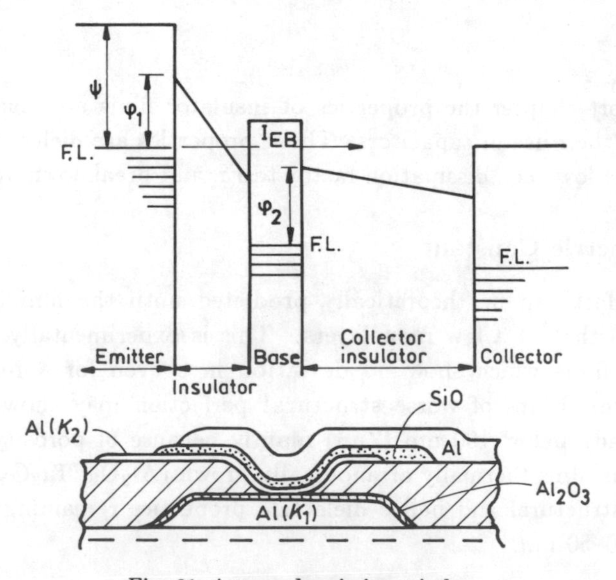


Fig. 61. A tunnel-emission triode.

EMITTER. The efficiency is, however, very low. Emission current = $10^{-4}$ – $10^{-3}$ total current. Using the tunnel emitter for injecting electrons into the conduction band of a further insulator a TUNNEL-EMISSION TRIODE (Fig. 61) is set up. Unfortunately the triode suffers from all the problems of the tunnel emitter plus ones arising from the second MIM part. The main difficulty is caused by the requirement of very thin perfect films.

# CHAPTER 5

# DIELECTRIC PROPERTIES OF THIN INSULATOR FILMS[10]

In this short chapter the properties of insulator films are discussed with respect to their use in capacitors. These properties are dielectric constant $\varepsilon$, dielectric loss, i.e., dissipation factor $\tan\delta$, and breakdown voltage $U_D$.

## 5.1. Dielectric Constant

No size effect can be theoretically predicted until the film thickness is reduced to that of a few monolayers. This is experimentally verified for Langmuir films which show no deviation in $\varepsilon$ even for a monolayer of Cd-stearate. Films of worse structural perfection may show a decreasing $\varepsilon$ already below 100 nm (ZnS) mainly because of porosity (Fig. 62). Amorphous films thermally or anodically grown ($Al_2O_3$, $Ta_2O_5$) also show excellent structural and hence dielectric properties remaining unchanged down to 10–50 nm.

## 5.2. Dielectric Loss

$\tan\delta$ = loss current/blind current = $1/\omega RC$ (Fig. 63), $\omega$ being the angular frequency of the current. The shunting effect of $R$ is frequency-dependent and, according to the particular material, rather complex dependency of $\tan\delta$ on $\omega$ may be observed. The thin film effect on the loss is mainly created by the difference in structure as compared to bulk dielectrics. For

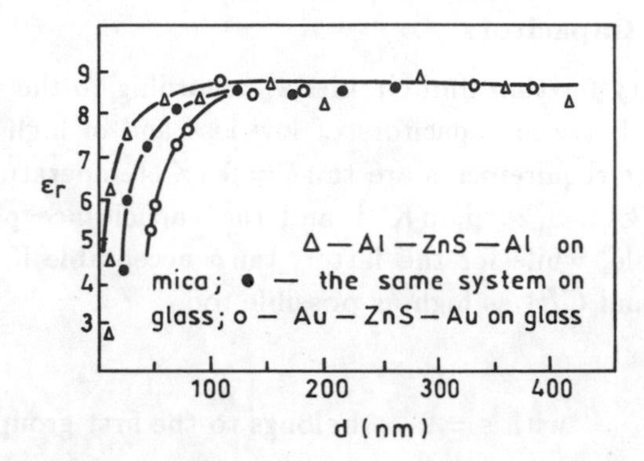


Fig. 62. Dielectric constants $\varepsilon_r$ vs. film thickness $d$.

the amorphous oxides $Al_2O_3$ and $Ta_2O_5$ fairly constant losses in the range of audio frequency have been observed, e.g., for $Ta_2O_5$, $\tan\delta \approx 4 \times 10^{-3}$.

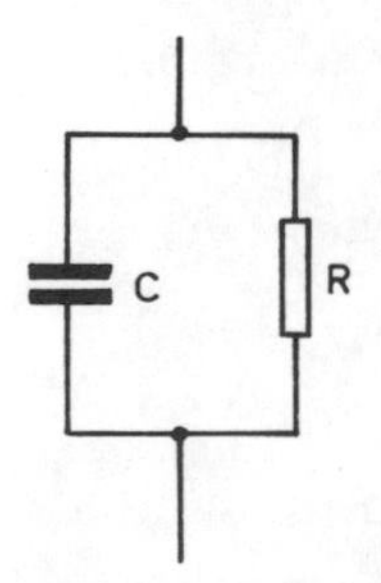


Fig. 63. RC circuit.

## 5.3. Breakdown Voltage

This voltage is usually somewhat higher for thin films than for bulk dielectrics, especially in thermally grown oxides. The absence of leak paths and the inhibition of avalanche if the thickness is small is usually responsible for this phenomenon. The best value of breakdown voltage is around 10 MV $cm^{-1}$. The most likely mechanism for a breakdown is thermal runaway. Near the breakdown voltage, small regions begin to draw a higher current thus heating this region which, because of TCR $< 0$, consequently enhances the current, the heat, etc.

## 5.4. Thin Film Capacitors

They are necessary for thin film circuitry. According to the requirements one distinguishes between capacitors of low loss and of high capacitance. For the former the requirements are $\tan\delta = 0.1\%$, temperature coefficient of capacitance TCC $< 20$ ppm $\mathrm{K}^{-1}$ and the capacitance per area $C/A$ as high as possible, while for the latter, $\tan\delta$ acceptable if $= 1\%$, TCC $< 500$ ppm $\mathrm{K}^{-1}$ and $C/A$ as high as possible too.

$Al_2O_3$ with $\varepsilon = 9$ belongs to the first group

$SiO$ $\varepsilon = 6$ (protected)

$SiO_2$ $\varepsilon = 4$

$Ta_2O_5$ $\varepsilon = 26$, to the second group.

# CHAPTER 6

# SUPERCONDUCTING FILMS[11]

Before discussing specific thin film phenomena in superconductivity we shall briefly review its general features.

## 6.1. General Features

Superconductor has now been detected already in some thousands of different metals and alloys. The most striking characteristic of superconductivity is the breakdown of its electrical resistance below the transition temperature, usually in the region below some K. Superconductors expel magnetic flux (resulting either from external fields or from its own current) from its interior to a small surface layer (about 400 A), where it decreases exponentially with penetration distance. Hence the current-carrying layer is of the same extension. When the field strength exceeds a critical value superconductivity cannot be maintained for a certain type of material (type I or soft superconductors, all superconductive elements except Ti and V) whereas type II or hard superconductors start to be penetrated by the flux in special tubes (vortices) showing normal state besides superconductive regions. For type II materials superconductivity is completely lost above the so-called upper critical field.

Microscopically, at the superconductivity transition the conduction electrons start to condense to pairs which have a lower energy than the two single ones (Cooper pairs). The coupling mechanism is the mutual lattice deformation (exchange of virtual phonons) and extends over $10^{-4}$ cm. The coupled electrons show opposite spins and momenta. At $T = 0$ all electrons are assumed to be condensed to Cooper pairs. At $T > 0$, the bond may be broken and an energy of some meV is taken from the thermal reservoir. The binding energy appears as an energy gap around the Fermi level. The recently found high-$T_c$ ceramic superconductors may have completely different mechanisms of superconductivity which is unclear up to now.

## 6.2. Thin Film Effects

It is difficult to account for all the single effects which make thin film superconductors different from their bulk counterparts. Experiments have shown that difference in critical temperature of the order a degrees is possible. With decreasing film thickness some materials show increasing $T_c$ (e.g., In and Al) whereas for others a decrease (Pb) or a maximum (Sn, Tl) has been observed. In some cases such effects can be correlated with internal stress induced during growth or by incompatible substrates. Depending on their special structure films may show a much higher $T_c$ . For example, Al has $T_c = 1.2$ K and 4.2 K for bulk and film sample respectively, while if Bi and Be, which are not bulk superconductors, are deposited as amorphous films they exhibit superconductivity with $T_c = 6$ K and 8 K respectively. The effect of impurities in low concentration is generally a linear decrease of $T_c$ with concentration according to $T_c = -Kc$. Hence superconductive film with a higher getter ability for oxygen such as V, Ta and Nb may show considerable deviation from bulk superconductors. ($K = 0.5$ to 1.3 Kat%$^{-1}$).

The size effect on the critical field depends mainly on the type of superconductor. It increases with decreasing thickness, e.g., for Type I superconductors, the parallel critical field is given by

$$H_{\mathrm{cf}} = H_{\mathrm{cb}}\sqrt{6}\lambda/D \tag{67}$$

for $D < \lambda$, whereas for $D > \lambda$ the expression

$$H_{\mathrm{cf}} = H_{\mathrm{cb}}\,(1 + \lambda/D) \tag{68}$$

is valid. For a perpendicular field, theory predicts

$$H_{\mathrm{cf}} = H_{\mathrm{cb}}\,D/b \tag{69}$$

where $b$ is width of the film and the subscripts f and b denote the film or the bulk respectively. For type II superconductors the treatment is more complex and must account for the influence of D on the Ginzburg–Landau parameter as well. The critical current $I_c$ is the one which generates the critical field at the surface. For a thin film wound to form a cylinder of radius $a$ Silsbee's rule states

$$I_c = H_c \, 2\pi a$$

As the current density is decreasing with surface distance by $\exp(-x/\lambda)$, the critical value is given by $j_c = H_{c/\lambda}$ which must not be exceeded. If $D \ll \lambda$ then the critical current is

$$I_{cf} \approx j_c Db = H_c bD/\lambda \tag{70}$$

If $D \ll \lambda$ it is

$$I_{cf} = H_c \, b \tanh D/2\lambda \tag{71}$$

For high-$T_c$ superconductors a thin film usually has much higher critical current than the bulk. The reason is unknown up to now.

## 6.3. Applications, Thin Film Cryogenic Devices

The most attention so far has been paid to memory/storage devices utilizing switching between normal and superconductive states. High speed elements showing switching times up to 50 ps may be produced, promising computers 10 to 100 times faster than those currently available. They also use a power level $10^4$ smaller than at present. An estimate shows that it should be possible to produce processors of 1 kbits/cm$^3$ with 1 W refrigeration capacity.

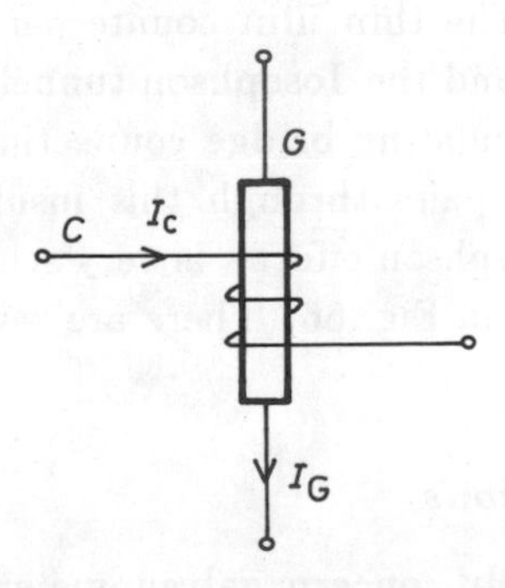


Fig. 64. A bulk version of cryotron.

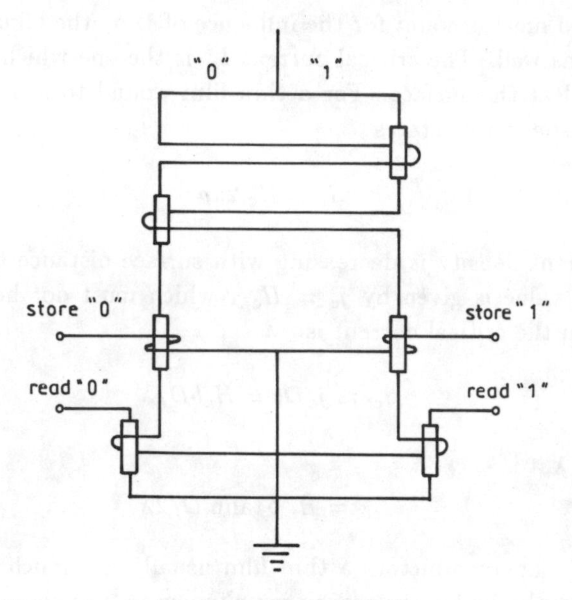


Fig. 65. The 1 bit memory device consisting of a flip-flop, "write" and "read" units.

### 6.3.1. *Switching elements, cryotrons*

In Fig. 64 the bulk version of the cryotron is shown. The control coil produces a field H higher than the critical field of the gate, thus switching it from superconductivity to normal state. The maximum gate current is $I_{\mathrm{G\,max}} = D\pi H_{\mathrm{cG}}$ while the control current is $I_{\mathrm{c}} = H_{\mathrm{cG}}/n$, $n$ being the number of windings. It is necessary to have $I_{\mathrm{G}}/I_{\mathrm{c}} = \pi Dn > 1$ to enable a cryotron to switch others. Logical units may be set up by several cryotrons e.g., 1 bit memory device basically consisting of a flip-flop and "write" and "read" units (Fig. 65). The thin film counterparts of bulk cryotron are the crossed film cryotron and the Josephson tunnel cryotron which utilizes the switching of a thin insulating bridge connecting two superconductors. The tunneling of Cooper pairs through this insulator, thus becoming a superconductor itself (Josephson effect), is very sensitive to magnetic field. Both devices are depicted in Fig. 66. There are several other logic devices as well.

### 6.3.2. *Further applications*

Further applications mainly concern galvanometers and fluxmeters. The principle on which such devices are based are Super-Conducting-Quantum-

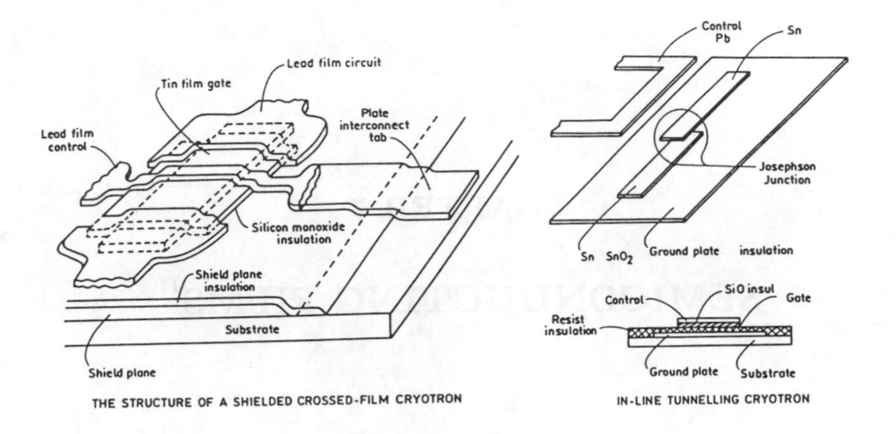


**Fig. 66. In-line tunneling cryotron using thin films.**

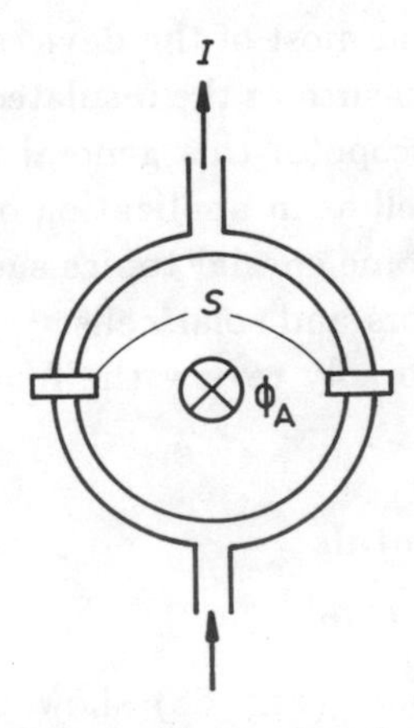


**Fig. 67. Interference magnetometer with two Josephson junctions. S–junctions.**

Interference-Devices (SQUIDS)(Fig. 67). The sensitivity of the Josephson tunneling current on an external magnetic field is enhanced by inserting two junctions into a superconducting loop. The electron wave functions corresponding to the currents flowing through both junctions interfere, thus defining the total current. Changes in the magnetic flux in the loop affect the phases of the wavefunctions, thus becoming detectable down to $10^{-11}$ G. The AC Josephson effect can be used for microwave generation from 1 to 1000 GHz with a spectral purity of about $10^{-7}$. Spectral breadth is given by the temperature, which is thus measurable down to some mK.

# CHAPTER 7

# SEMICONDUCTING FILMS[11]

Modern semiconductor device fabrication involves many thin film technological production steps and most of the devices include thin films as parts essential for their operation such as the insulated gate field effect transistor. It would be beyond the scope of this general textbook to discuss all the details in technology as well as in application of semiconductors. We shall therefore deal only with some special topics such as devices for integrated circuits, thin film transistors and solar cells.

First, however, let us briefly review the fundamental features of active semiconductor devices.

## 7.1. Physical Fundamentals

### 7.1.1. *Electrical conduction*

The simplified band schemes (Fig. 68) show that at zero temperature a semiconductor exhibits a filled valence and an empty conduction band. For elevated temperatures electrons may be emitted into the conduction band. The number is ruled by a probability factor $\exp(E_g/2kT)$ for the intrinsic semiconductor. The same amount of empty states (holes) is left in the valence band (pair formation). Impurities showing a valence electron in excess which does not contribute to the chemical bond (donors) are easily ionized, thus emitting their excess electrons into the conduction band ($n$-type semiconductor). On the contrary, impurities with one valence electron in deficit may be ionized by taking a thermally emitted electron from the valence band leaving behind a mobile hole ($p$-type semiconductor). At room temperature all such impurities are ionized since the level of their electron states does not differ much from the bottom of the conduction band or the top of the valence band.

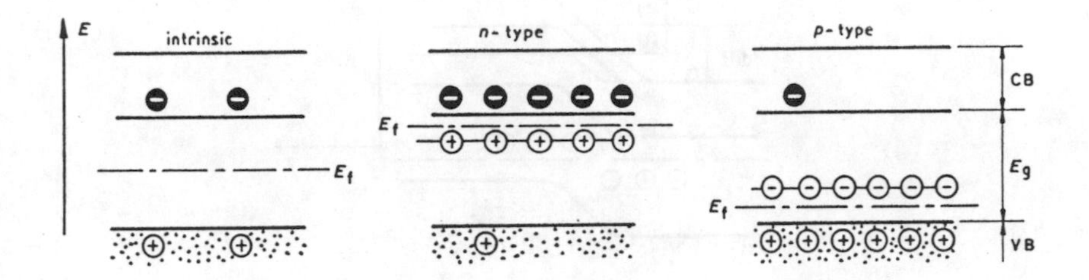


Fig. 68. The simplified band scheme of a semiconductor.

The concentrations of holes, $p$, and electrons, $n$, are governed by the law of mass action: $np = \text{const}$. If one or both types of charge carrier are present in excess they will be reduced by recombination. In practice the equilibrium concentration of electrons is equal to the concentration of donors, $n_D$. Hence the holes in a $n$-semiconductor are present in a concentration $p_D = \text{const.}/n_D$, much lower than in the intrinsic case. The same holds for holes in a $p$-semiconductor. The change from non-equilibrium to equilibrium is given by the lifetime $\tau$ of the minority charge carriers (time needed for finding a partner for recombination). During their life time they may diffuse as a result of their statistical thermal movement. The average length traversed is called diffusion length, $L_e$ and $L_h$ for electron and hole respectively.

### 7.1.2. *p-n junctions* (Fig. 69)

Without an external electrical field the mobile charge carriers diffuse through the junction of the $p$-$n$ diode into the opposite type semiconductor where they recombine, thus creating a non-mobile space charge associated with a contact voltage (diffusion voltage), which finally undergoes further diffusion. The junction is depleted of mobile carriers. An external voltage positive on $n$- and negative on $p$-semiconductor supports the diffusion voltage, consequently expanding the depletion zone and inhibiting a flux of carriers by diffusion. The junction is non-conducting (blocking direction). If the bias is of opposite polarity, then the diffusion voltage is partly compensated and a steadily flowing diffusion current tries to re-establish

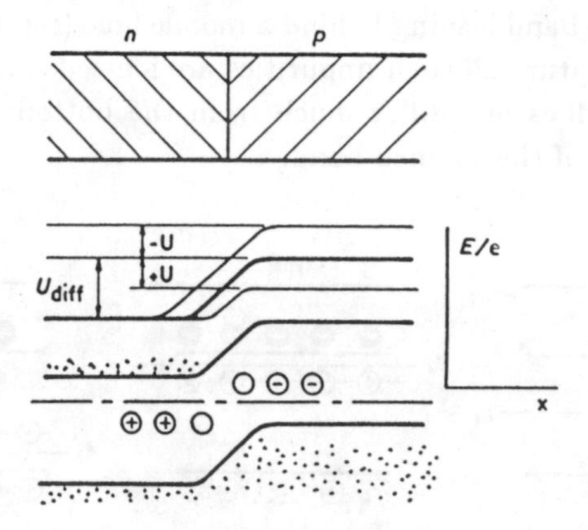


Fig. 69. A *p-n* junction.

equilibrium. The junction becomes conducting. The diffusion current of minority carriers changes finally to a flux of majority carriers by recombination (forward direction).

The bipolar transistor (Fig. 70) is an array of two *p-n* junctions either in *pnp* or in *npn* sequence. The first is biased in forward direction (emitter diode), while the second in blocking direction (collector diode). If the sandwiched semiconductor (base) is thinner than the diffusion length the minority carriers may penetrate it, thus arriving at the collector junction through which they can penetrate. The flux of these (minority) carriers is a diffusion flux ruled by the concentration gradient which is controlled by the biases of EB (emitter-base) and BC (base-collector) diodes, thus enabling amplification of electric signals.

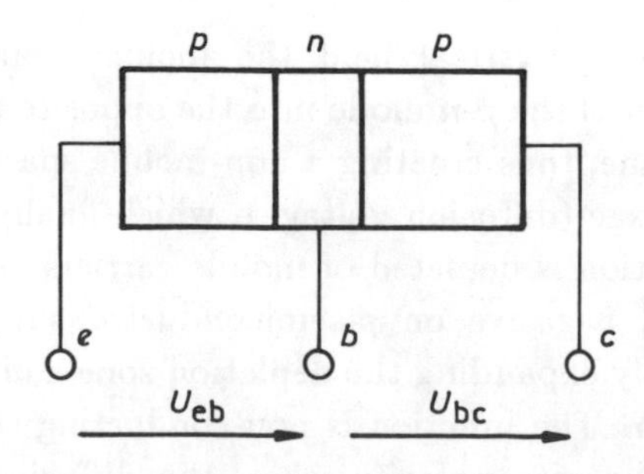


Fig. 70. A bipolar transistor.

### 7.1.3. *Field effect*

A completely different principle is utilized by field effect transistors, of which we shall discuss the Insulated Gate Field Effect Transistor (FET). This type may be successfully produced as a pure thin film device, to be discussed later. In Fig. 71 the band scheme of a *p*-Si-$SiO_2$-Al structure is shown. The $SiO_2$-Si interface is depleted of holes by the band bending. If the Al electrode is positively biased, the Si conduction band is further bended, finally overlapping the Fermi level. Hence it will be filled by electrons (via pair formation) thus providing a *n*-conducting (inversion) channel next to the surface. Its cross section is given by the bias. Instead of the thermal filling the channel can be filled by electrons injected from a *n*-conducting region (source). A *n*-region can also be used for draining an electron current through the channel (drain). Its amount can be controlled by the insulated metal electrode (gate), hence allowing amplification and switching. Practically because of the surface states a Si surface is usually of *n*-type so that inversion in *p*-Si can be already present at zero bias. Such FET is called self-conducting depletion type, whereas non-conducting at zero voltage is called enhancement type FET.

Let us now discuss how such elements are manufactured.

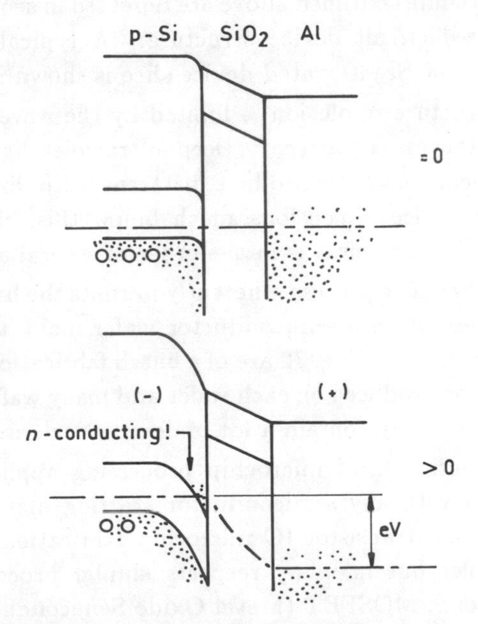


Fig. 71. The band scheme of a p-Si-$SiO_2$-Al structure.

## 7.2. Devices for Integrated Circuits (IC)[12]

The minute device structures of integrated circuits are delineated on the semiconductor wafer surface with the aid of photolithographic techniques. The first step in the device fabrication procedure for Si ICs is to coat an oxidized epitaxial slice of Si with a liquid photoresist. Then the whole coated slice is placed under a masking plate and exposed to ultraviolet radiation. After washing with a solvent (developed) the desired photoresist pattern is left on the oxide. The oxide is then selectively removed from certain areas and allowed to remain in others by subjecting the slice to a buffered solution of HF acid which dissolves $SiO_2$ where it is not protected by the photoresist. In this way an oxide mask is produced which will later inhibit selectively subsequent diffusion of such doping impurities as B and P when the slice is subjected to a vapor of these impurities at a temperature in excess of 1000°C. A metallic interconnection pattern is also similarly delineated by depositing the metal everywhere on the slice using photoresist for protection and an appropriate acid to dissolve the metal where unwanted. Metals commonly used are Al and refractory metals. Polycrystalline Si and metal silicides are also used for device interconnections.

The basic procedures outlined above are repeated in sequence to produce the final integrated circuit device structure. A typical set of steps to fabricate a monolithic Si integrated device slice is shown in Fig. 72.

The device structure resolution is limited by the wavelength $\lambda$ of light used to project the mask pattern. Deep ultraviolet light ($\lambda$ = 0.2–0.3 micrometer) is being used to produce patterns with line widths of 0.5 micrometer. For device dimensions much below this, electron or X-ray irradiation is used since their $\lambda$ is in the order of several angstroms.

The photolithographic process inherently permits the basic micro-circuit pattern to be repeated on a semiconductor wafer many times. All of the processing steps outlined in Fig. 72 are of a batch fabrication type. Literally many circuits can be produced on each wafer and many wafers are processed at once. Each circuit may contain a lot of devices such as transistors.

The above description of microchip processing applies to integrated circuits fabricated with any semiconductor starting material. A specific process for Si bipolar transistor IC microchip fabrication is illustrated in Fig. 72. A simpler but in many respects similar procedure is used in the manufacture of Si MOSFET (Metal Oxide Semiconductor Field Effect Transistor) ICs. GaAs microchip digital ICs and optically coupled ICs

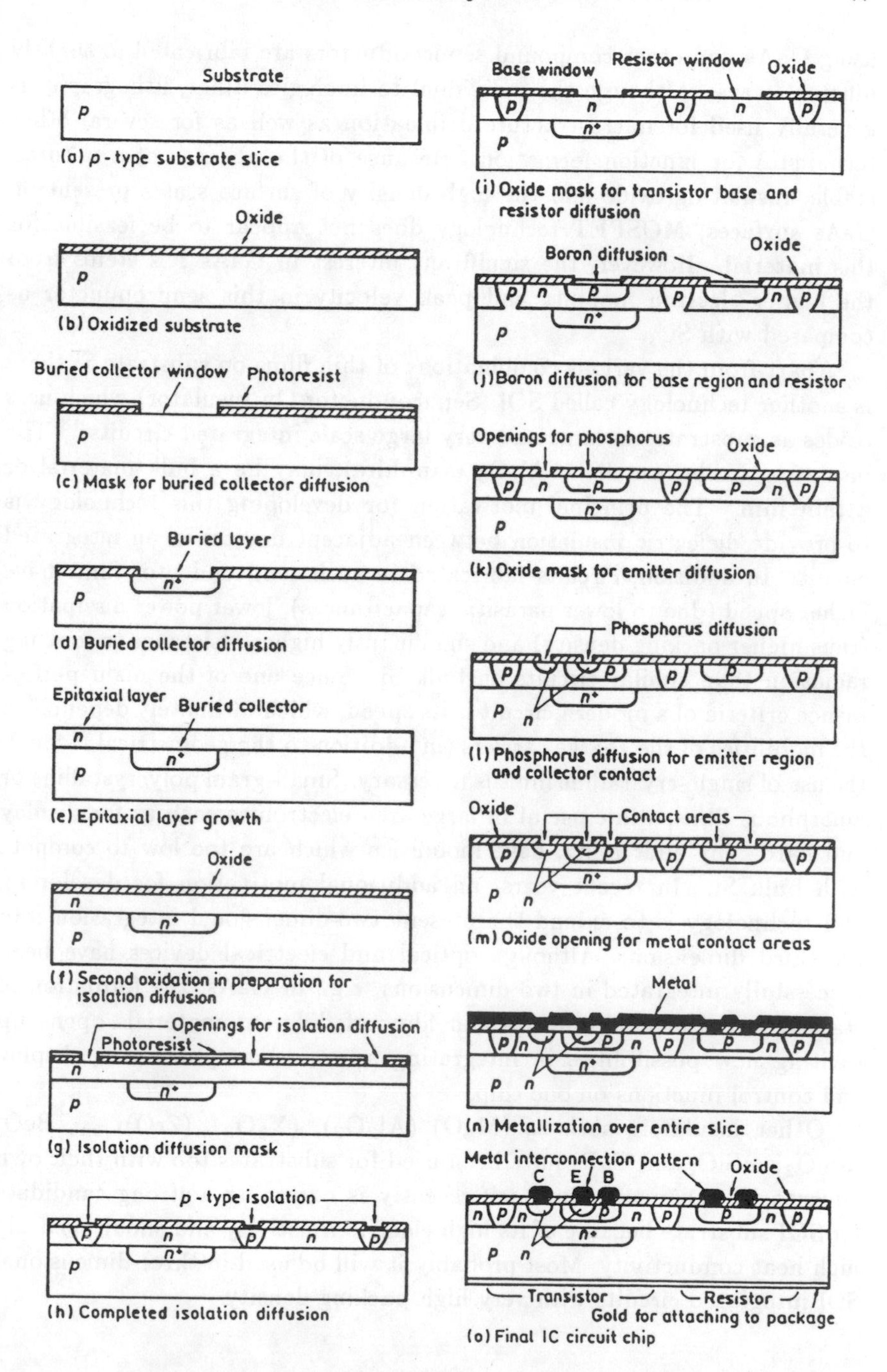


Fig. 72. The production of a typical IC (integrated circuit).

using GaAs and other compound semiconductors are fabricated in slightly different ways. Although the individual technologies differ, lithography is generally used for microstructure delineation as well as for several other techniques for junction formation. Because of the absence of a natural, stable, insulating oxide and the high density of surface states present on GaAs surfaces, MOSFET technology does not appear to be feasible for this material. However, the significant interest in GaAs ICs stems from the higher electron mobility and peak velocity in this semiconductor as compared with Si.

Apart from the various combinations of thin films on substrate Si there is another technology called SOI (Semiconductor On Insulator) which uses oxides as substrates in large and very large scale integrated circuits.[13] The insulating oxide, usually $\alpha$-$Al_2O_3$ (sapphire), may be a bulk material or a thin film. The principal motivation for developing this technology is to provide dielectric insulation between adjacent devices in an integrated circuit. In addition, circuits fabricated in such semiconductor films have higher speed (due to lower parasitic capacitances), lower power dissipation (thus higher packing density) and significantly higher resistance to ionizing radiation than similar circuits on bulk Si. Since one of the main performance criteria of a modern circuit is its speed, which ultimately depends on the mobilities of the charge carriers (in addition to the geometrical layout), the use of single-crystalline films is necessary. Small-grain polycrystalline or amorphous films, while useful in large-area electronics such as for display and hard-copy addressing, have mobilities which are too low to compete with bulk Si. In recent years, an additional motivation for developing SOI technology is to extend the present two-dimensional integration into the third dimension. Although optical and electrical devices have been successfully integrated in two dimensions, e.g., in GaAs, the formation of stacked layers of device-worth thin films of different materials opens up exciting new possibilities in integrating sensor, data processing, display and control functions on one chip.

Other insulators such as (MgO) ($Al_2O_3$), $(Y_2O_3)_m(ZrO)_{1-m}$, BeO, $\alpha$-$SiO_2$, $\alpha$-SiC, and $CaF_2$ have been used for substrates too with their own advantages. Diamond film itself recently is becoming a strong candidate for SOI substrate because of its high electric resistivity and simultaneously high heat conductivity. Most probably it will be used in three-dimensional SOI integrated circuits with very high packing density.

## 7.3. The Thin Film Transistors (Fig. 73)

The Thin Film Transistors (TFT) is essentially a field effect transistor, since bipolar ones are confined to single crystals. A glass substrate supports a semiconductor film, usually CdS or CdSe, deposited between two metal strips predeposited to form source and drain. The semiconductor is covered by a dielectric (e.g., SiO) on which finally the gate electrode is deposited. Such elements have been successfully used as switches which can be integrated to form shift registers for matrix-addressed display panels, thus showing the advantage of saving external interconnections. Such arrays have been produced in one single pump down by means of movable wire grill masks.

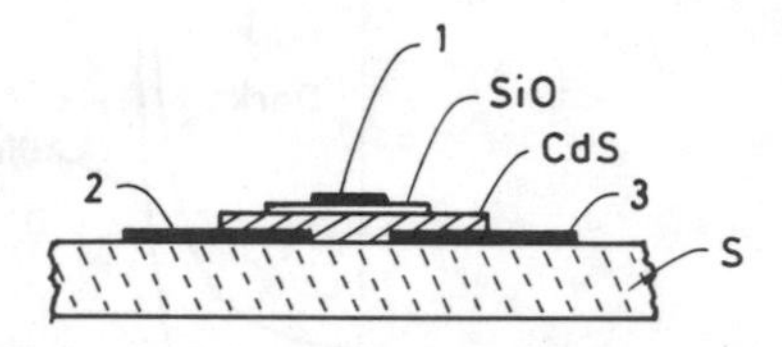


Fig. 73. A thin film transistor (TFT).

## 7.4. Photovoltaic Cells

Interest in solar energy-converting elements results mainly from requirements in the space program but has now received great attention because of the depletion of fossil energy resources. Direct conversion of radiation into electric energy is certainly more desirable than that via heat but has up to now not yet reached a significant economic standard. Nevertheless the technology appears to be promising and for a physicist its principle is in any case of intrinsic interest.

### 7.4.1. *Principle*

Let $\alpha$ be the linear absorption coefficient of radiation in a semiconductor. Then the absorbed intensity per unit volume is

$$I = I_0(1 - e^{-\alpha x}) \approx I_0 \alpha x \qquad (\alpha x \ll 1) \tag{72}$$

where $I_0$ is intensity of incident radiation, $x$ the penetration distance. The number of photons per volume is then $I_0\alpha/h\nu$, which causes an increase of hole-electron pairs above the equilibrium level at a rate $R_G = I_0\alpha q/h\nu$ ($q$

is quantum efficiency). The charge carriers so created increase the current density of a *p-n* junction biases in blocking direction by an amount

$$j_{\mathrm{L}} = \frac{I_0 \alpha q}{h\nu} e(L_{\mathrm{e}} + L_{\mathrm{h}}) \propto I_0 \, . \tag{73}$$

Hence we obtain the diode characteristics for illumination by shifting those for the dark by $j_{\mathrm{L}}$ (Fig. 74). A resistor connected in series to a diode and an (ideal) voltage source receives a voltage drop according to the intersections of its own and the characteristic of the diode. For zero external voltage, similarly, the voltage on the resistor and the respective current are given by the intersections of its characteristic with the abscissa.

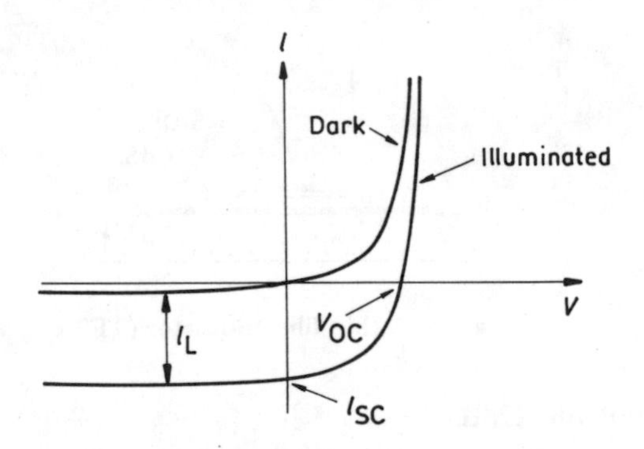


Fig. 74. The *I-V* curves of a diode photocell.

Analytically the characteristics are given by

$$j = j_{\mathrm{D}} - j_{\mathrm{L}} = j_0 e^{Ve/kT} - \frac{I_0 \alpha q}{h\nu} e(L_{\mathrm{e}} + L_{\mathrm{n}}) \, , \tag{74}$$

yielding for $j = 0$ the open circuit voltage

$$V_{\mathrm{oc}} = \frac{kT}{e} \ln \left[ \frac{I_0 \alpha q}{h\nu} \frac{e}{j_0} (L_{\mathrm{e}} + L_{\mathrm{n}}) \right] \, . \tag{75}$$

### 7.4.2. *Practical aspects*

Hetero-junctions as e.g., CdS-$Cu_2S$ film junction is of *p*-($Cu_2S$), *i*-(mixed Cu-, CdS), *n*-(CdS) type. Its power conversion efficiency can be improved by heat treatment. Best values are around 7–8%. Open circuit voltages are about $V_{\mathrm{oc}} = 0.5$ V, short circuit current densities $j_{\mathrm{sc}} = 30$ mAcm$^{-2}$.

Other hetero-junctions are InP-CdS, $InCuSe_2$-CdS, GaAs-ZnSe, CdTe-CdS. The most commonly used homo-junction is the Si homo- junction prepared by CVD epitaxial growth. Polycrystalline thin film Si homojunctions have also been investigated: $V_{oc} = 0.6$ V, $j_{sc} = 20$ $mAcm^{-2}$. Power conversion efficiency 5%.

Amorphous Si (*a*-Si) thin films have recently been developed as a cheap material for large-scale applications of photovoltaic cells.[14] Most of them are used to power pocket calculators and to charge the battery of electronic watches. Larger *a*-Si cells are now becoming commercially available for battery charging, running portable radio sets, etc. The rate of development in this field in recent years has been impressive. In 1976 the early junctions had a conversion efficiency of about 2%. By 1982 it had been increased to 10% over small areas (1 cm). A major achievement is the large-area cells (10 cm × 10 cm) which now, in production, possess efficiencies between 7 and 8%. In recent years several interesting technological advances have been made, improving the performance and stability of the original *p-i-n* devices. For instance, a thin glow-discharge Si-C layer on the *p*-side produces a wide-gap window effect which increases the built-in potential to over 0.9 V. Another type of hetero-junction involves the use of a thin surface layer of glow-discharge $\mu$c-Si to form a ITO/$n$-$\mu$cSi/$i$/$p^+$/$TiO_2$-Ag/SS structure. The $TiO_2$/Ag electrode contributes to the more efficient collection of the weakly absorbed longer-wavelength component. This increases the short-circuit current by 20% and leads to a value of conversion efficiency equal to 9.2%. The open circuit voltage reaches 0.88 V and the short-circuit current density, 16.1 $mA/cm^2$. Promising results have also been obtained with tandem-type cells in which layers having different band gaps are stacked in the device. Possible materials are *a*-Si-N, *a*-Si and *a*-Si-Ge films, all produced by glow-discharge decomposition of suitable gas mixtures. It is estimated that the economic use of *a*-Si photovoltaic devices in large-scale power applications requires 8–10% conversion efficiencies over large areas. Several solar houses and roof-top test arrays are in operation and the application of *a*-Si cells in third-world irrigation systems appears a possibility in the near future.

### 7.4.3. *Future aspects*

The maximum insolation on earth is around 1 $kWm^{-2}$. This gives a total power of 100 kW from 1 $km^2$ if a future efficiency of 10% is assumed. Hence, a small part of the Sahara desert could in principle supply all the electric power for Europe, the Middle East and Africa. The main problem, however, is the daily and annual variations of insolation. Just when a higher energy consumption is required there is only little insolation (evenings, winters). Hence proper storage must be assured, which up to now is not yet solved too. Several feasibility studies have been carried out, which give the conditions for solar energy conversion to be competitive to conventional energy providing systems. For large-scale use it seems necessary that the array of cells should have an efficiency $>10\%$ maintained over a period of 20 years, and the material of which solar cells are made should be available in sufficient quantities as that production is not limited by depletion of resources.

# PART II

# CHAPTER 8

# THIN FERROMAGNETIC FILMS[15]

## 8.1. General

In all cases where miniaturization-dense integration is a major requirement for a device one is first thinking of thin film technological solutions of this problem. Speaking in terms of information storage thin magnetic films appear to be suitable not only for thin film ring cores. Having higher integration densities in mind, planar devices allowing for ferromagnetic switching would be even more convenient. Both attempts to introduce thin films into the magnetic storage began to be fruitful in the late fifties. Meanwhile storage units based on principles other than magnetism reversal have been developed. These devices make use of special domain structures uniquely associated with the information to be stored, shifted and read out. In the following part we shall deal with all these possibilities, where films of soft magnetic materials play the dominant role. Here we deal mostly with films of "Permalloy"- NiFe 81/19 with a typical thickness around 1000 Å. However, we shall also briefly mention the application of hard magnetic films for permanent storage.

## 8.2. Fundamental

We bear in mind that ferromagnetism is given by the alignment of uncompensated $3d$ electron spins by quantum-mechanical exchange in the elements Fe, Co, Ni and their alloys, as well as others among which rare earth elements play an outstanding role. For our considerations we only need a few characteristics of a special material:

(1) The saturation magnetization $M_s$ given by the sum of the moments of perfectly aligned dipoles per unit volume.
(2) The anisotropy properties of the material in thin film form.
(3) The domain wall properties in thin films. In order to answer the question which orientation $M$ will take in a given sample under special circumstances one has to find the direction in which minimum potential energy is achieved (which consideration is equivalent to looking for the equilibrium of torques on the individual spins).

### 8.2.1. *Energy contributions (All energy terms are given for unit volume — equations only locally valid!)*

#### 8.2.1.1. *External field:* H

If for any reason the angle between **M** and **H** is not zero, then

$$E_{\mathrm{H}} = -\mathbf{M} \cdot \mathbf{H} \,. \tag{76}$$

#### 8.2.1.2. *Stray field, demagnetizing field*

In regions where $\nabla \cdot \mathbf{M} = 0$ (mainly at surfaces or other inhomogeneities) magnetic poles are created. Their field is called "demagnetizing field ($\mathbf{H}_{\mathrm{mag}}$)" inside and "stray field ($\mathbf{H}_s$)" outside the sample. The demagnetizing field acts on $M$ so that a potential energy

$$E_{\mathrm{mag}} = -1/2\mathbf{H}_{\mathrm{mag}} \cdot \mathbf{M} \tag{77}$$

arises. The factor 1/2 accounts for the fact that $\mathbf{H}_{\mathrm{mag}}$ is resulting from **M** itself: $\mathbf{H}_{\mathrm{mag}} = -N\mathbf{M}$, $N$ being the demagnetization factor. For an homogeneously magnetized ellipsoid $\mathbf{M} =$ const., poles are generated only at the surface such that $\mathbf{H}_{\mathrm{mag}} =$ const. as well. Thin films can be approximated by very flat disc-like ellipsoids as shown in Fig. 75.

$$N_{\parallel} = 4\pi t/(t+d) \ll N_{\perp} \approx 4\pi$$

hence one commonly finds $M$ within the film plane, where stray field is a minimum.

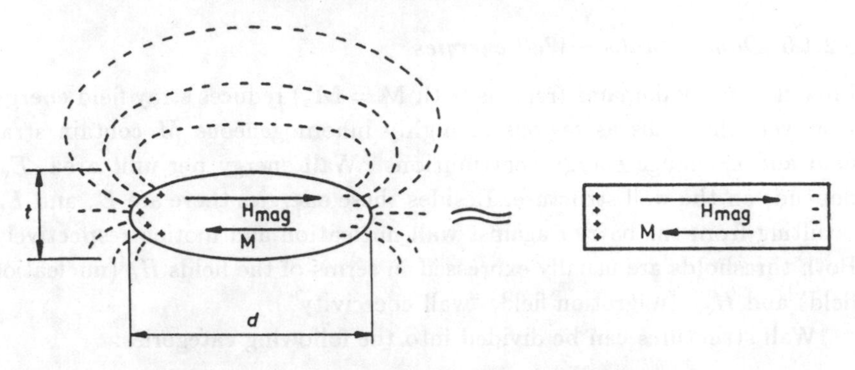


Fig. 75. The stray field in thin films. $N_{\|} \ll N_{\perp}$.

8.2.1.3. *Anisotropy*

Generally, even in spherical samples the potential energy may depend on the orientation of $M$ with respect to sample-fixed coordinates,

$$E_{\mathrm{a}} = K f(\alpha_1, \alpha_2, \alpha_3)$$

where $K$ is called "Anisotropy constant". The direction $\{\alpha_i\}$ for minimum $E_{\mathrm{a}}$ is called "Easy axis EA" while the direction for maximum $E_{\mathrm{a}}$ is "Hard axis HA".

The origin of anisotropy is twofold, one is the crystal anisotropy, e.g., **EA** is [100] in Fe and [111] in Ni, the other is the anisotropy induced by magnetostriction or by external field during solidification, growth or thermal treatment.

For binary switching in films UNIAXIAL anisotropy is important since

$$E_{\mathrm{a}} = K \sin^2 \theta \ ,$$

$\theta$ being the angle between **M** and **EA**.

A direct figure denoting the magnitude of anisotropy is the ANISOTROPY FIELD $H_{\mathrm{a}} = 2K/M_{\mathrm{s}}$, the field $\|$ **HD** which aligns **M** also in the **HD**.

8.2.1.4. *Exchange energy*

Minimum exchange energy makes $M$ perfectly aligned. Hence each deviation (expressed by $\nabla \cdot \mathbf{M}$) from parallel orientation enhances this energy,

$$E_{\mathrm{ex}} = \frac{A}{M} |\nabla \cdot \mathbf{M}|^2$$

$A$ is the "Exchange constant" characterizing magnetic stiffness.

### 8.2.1.5. *Domain walls – Wall energies*

Flux closure by domains (regions with $\mathbf{M} = \mathbf{M}_s$) reduces stray field energy. However, the walls as regions of highly inhomogeneous $M$ contain stray field and exchange energy contributions. Wall energy per unit area, $E_w$, depends on the wall structure. Besides these energies there are $E_n$ and $E_m$ resulting from the barrier against wall nucleation and motion respectively. Both thresholds are usually expressed in terms of the fields $H_n$ (nucleation field) and $H_{cw}$ (migration field, "wall coercivity").

Wall structures can be divided into the following categories:

(1) Bloch wall. **M** rotates ⊥ film plane, as shown in Fig. 76 which is a cross-sectional view of domain wall running normal to the plane of the page. There is a high stray field on the surface.

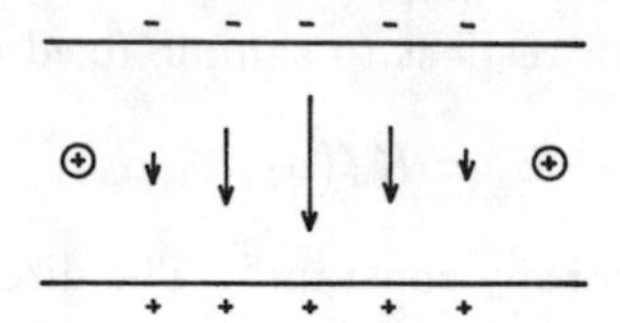

**Fig. 76. Bloch walls in thin films. Cross sectional view of domain wall structure. High stray field on surface.**

(2) Neel wall. **M** rotates in film plane, as shown in Fig. 77. There is a low stray field on the edges.

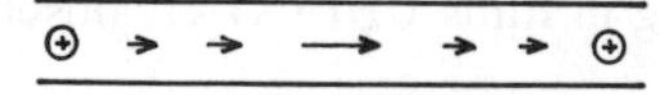

**Fig. 77. Néel wall. Cross sectional view of domain wall structure. Low stray field on edges.**

(3) Two-dimensional Bloch wall. **M** rotates in two directions as shown in Fig. 78. $\mathbf{M} = \mathbf{M}(x, z)$. Stray field is almost free on the surface.

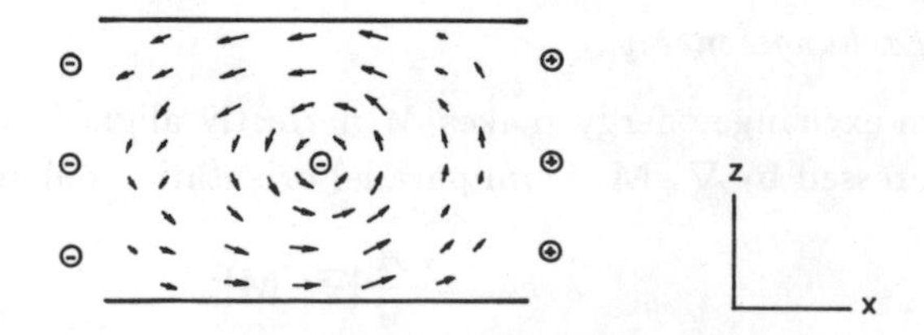


**Fig. 78. Two dimensional Bloch wall. Stray field free on the surface.**

(4) "Cross tie" wall. It may be considered as an array of Neel walls with different polarity separated by "Bloch lines" where **M** rotates out of film plane, as shown in Fig. 79. Spikes are Neel regions introduced in the regions where the curl direction would be opposite to $M$ of the adjacent domain.

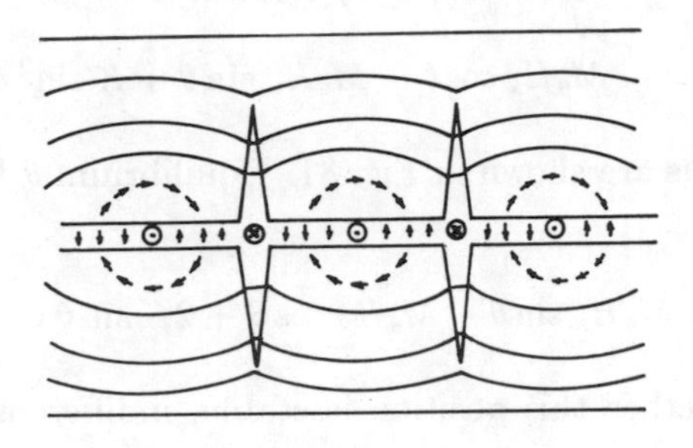

Fig. 79. "Cross-tie" wall.

It is obvious that in THIN FILMS walls are always perpendicular to the film plane to minimize wall energy.

The specific wall energy depends on film thickness, hence different types of walls are found in differently thick films. Figure 80 shows the theoretical curves of wall width and specific wall energy.

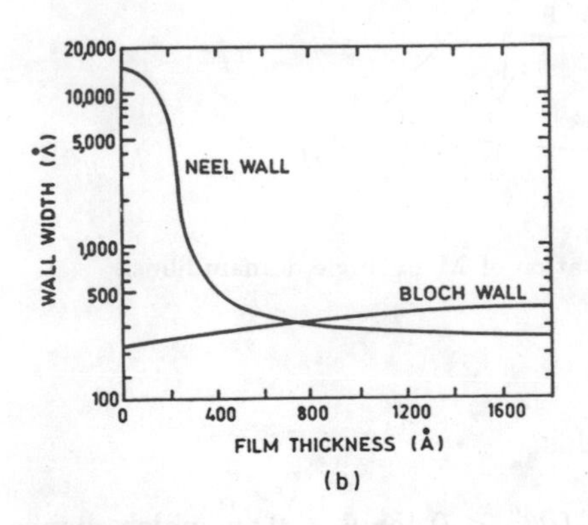


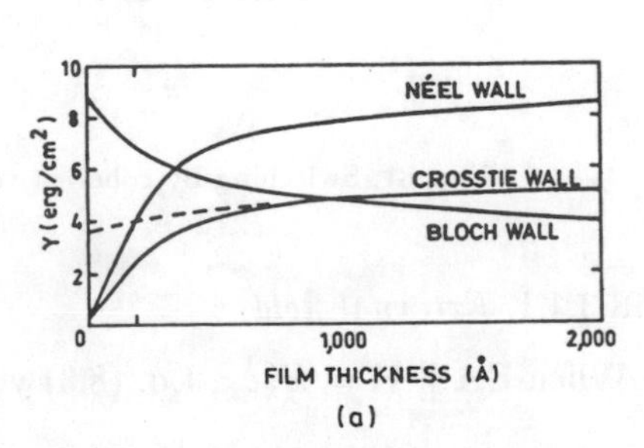


Fig. 80. (a) Theoretical surface energy density $\gamma$ of a domain wall as a function of thickness. The dashed curve shows the estimate of cross-tie-wall energy density when the energy of Bloch lines is included. (b) Theoretical widths of Bloch and Néel walls as a function of film thickness. [After S. Middelhoek, in J. Anderson (ed.), "The use of thin films in physical investigations", p. 385, Academic Press Inc., New York, 1966; and *J. Appl. Phys.* **34**, 1054 (1963).]

## 8.3. Magnetization Reversal in Uniaxial Thin Films

### 8.3.1. *Switching by coherent rotation of* M *in single domain films*

According to the Stoner Wohlfarth Model,

$$E = -M_s H \cos(\phi - \theta) + K \sin^2 \theta$$

$$= -M_s H_x \cos\theta - M_s H_y \sin\theta + K \sin^2 \theta$$

where the notations are shown in Fig. 81. Equilibrium $\theta$ for **M** is achieved where

$$\partial E/\partial\theta = M_s H_x \sin\theta - M_s H_y \cos\theta + 2K \sin\theta \cos\theta = 0 \,. \qquad (78)$$

The condition whether this position is stable, indifferent or unstable depends on the sign of

$$\partial^2 E/\partial\theta^2 = M_s H_x \cos\theta + M_s H_y \sin\theta + 2K(\cos^2\theta - \sin^2\theta) \,.$$

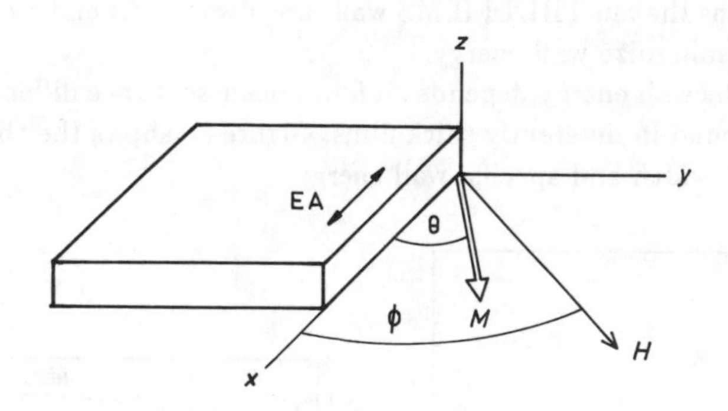


Fig. 81. Switching by coherent rotation of **M** in single domain films.

#### 8.3.1.1. *External field*

When **EA** $\parallel$ **H** $= H_x e_x$, Eq. (8.3) yields

(a) $H_x = -(2K/M_s)\cos\theta$ where $\partial^2 E/\partial\theta^2 < 0$ for $\theta \neq 0, \pi$ which, hence, describes unstable orientation.

(b) $\theta = 0$, $\partial^2 E/\partial\theta^2 = M_s H_x + 2K \gtrless 0$ if $H_x \gtrless -2K/M_s = H_k$
$\theta = \pi$, $\partial^2 E/\partial\theta^2 = -M_s H_x + 2K \gtrless 0$ if $H_x \lessgtr 2K/M_s = H_k$

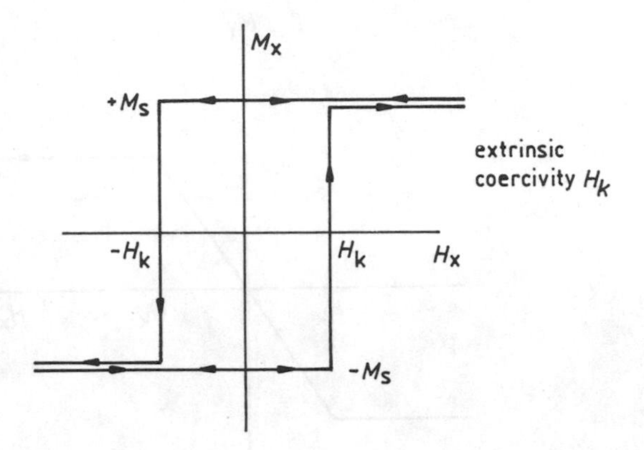


Fig. 82. Switching at the critical fields.

Two stable orientations $\theta = 0,\ \pi$ are possible for $-H_k < H_x < H_k$, outside which only the one coinciding with the orientation of $H$ is stable. Switching occurs at the critical fields $\pm H_k$ by coherent 180° rotation of $\mathbf{M}$ in the entire film. Extrinsic coercivity $H_k$ is observed. See Fig. 82.

8.3.1.2. *External field*

When $\mathbf{HA} \parallel \mathbf{H} = H_y e_y$, $\partial E/\partial\theta = 0$ for

(a) $\sin\theta = H_y/H_k$ where $\partial^2 E/\partial\theta^2 > 0$ for $\theta \neq \pm\pi/2$.
(b) $\theta = \pm\pi/2$ where $\partial^2 E/\partial\theta^2 > 0$ only for $H_y \gtrless \pm H_k$

Each $\theta$ is stable, i.e., for $\mathbf{H}$ rising from 0, $\mathbf{M}$ rotates towards $\mathbf{H}$ where complete alignment is achieved for $H \geq H_k$. No extrinsic coercivity is observed. See Fig. 83.

8.3.1.3. *Arbitrary direction of* $\mathbf{H}$

For a given $\theta$ all field vectors $\mathbf{H}$ point at a straight line defined by the magnetic energy (78). This orientation $\theta$ becomes unstable for the $\mathbf{H}$ for which $\partial^2 E/\partial\theta^2 = 0$. The envelope of (78) obeys

$$H_x^{2/3} + H_y^{2/3} = H_k^{2/3}\ ,$$

whose plot is called Stoner Wohlfarth astroid or Switching Curve, as shown in Fig. 84. It can be conveniently used to define $\mathbf{M}$ for any given $\mathbf{H}$ as $M_s$ is oriented parallel to the tangent from the tip of $\mathbf{H}$ on the astroid.

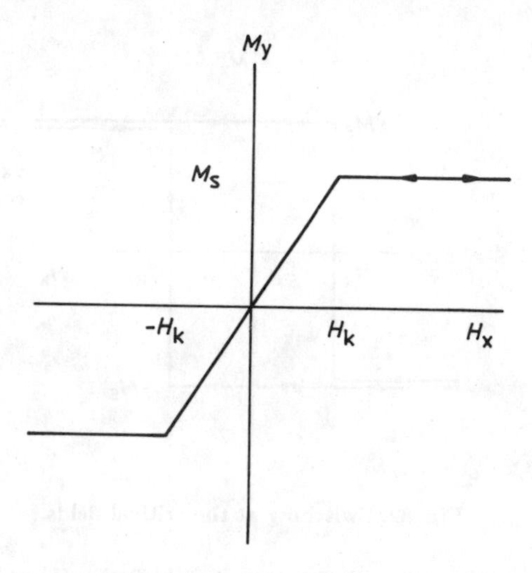


**Fig. 83. Each position is stable. No extrinsic coercivity.**

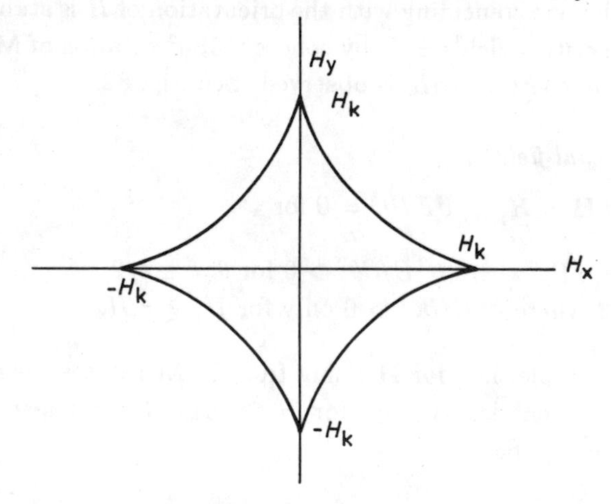


**Fig. 84. Stoner Wohlfarth Asteroid – Switching curve.**

This yields two stable directions of **M** when **H** is inside the switching curve and no switching can be achieved as long as **H** is altered such that the switching curve is not trespassed. There is only one stable direction of **M** for **H** outside the switching curve. Switching by the rotation of H and the increase of **H** are shown in Fig. 85 and 86 respectively. If $|H| > H_k$ then by rotation of **H** no switching but a continuous rotation of **M** is induced ($M_s$ is not always parallel to **H**!).

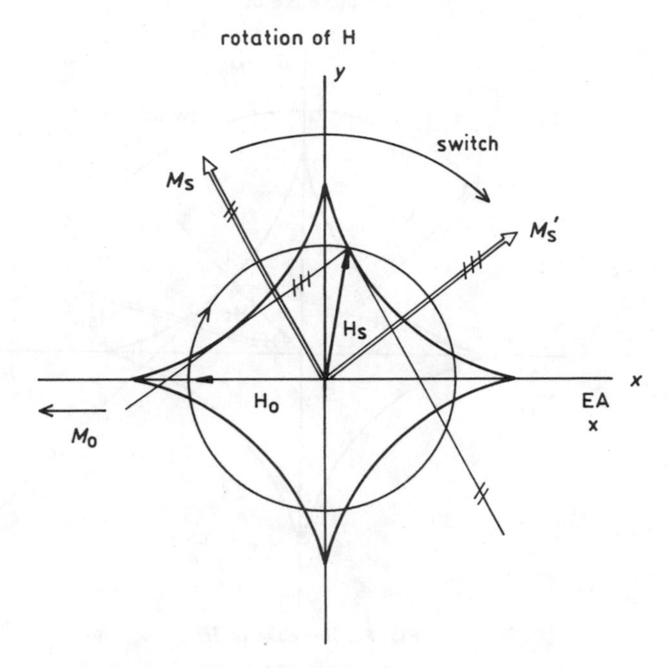


Fig. 85. Rotation of **H**.

### 8.3.2. *Reversal of* M *by domain growth*

Even if a film was ordinally free of domains the coercivity observed by applying EA-fields is sometime $< H_k$ and no flux components parallel to **HA** can be detected during switching.

By Lorentz electron microscopy the following mechanism for reversal was observed. After a threshold field $H_{cw}$ is exceeded domains of opposite magnetization migrate in the film until **M** is completely reversed. It is shown schematically in Fig. 87.

### 8.3.3. *Reversal by partial incoherent rotation and domain motion*

Magnetization ripples occur when the local perturbations cause **M** to fluctuate locally around the mean orientation parallel to the easy axis. A rising anti-parallel field hence causes partial clockwise or anticlockwise rotation of **M**, thus generating real domains as **H** is increased. When the thresholds for nucleation and wall motion are exceeded the reversal is completed by domain growth.

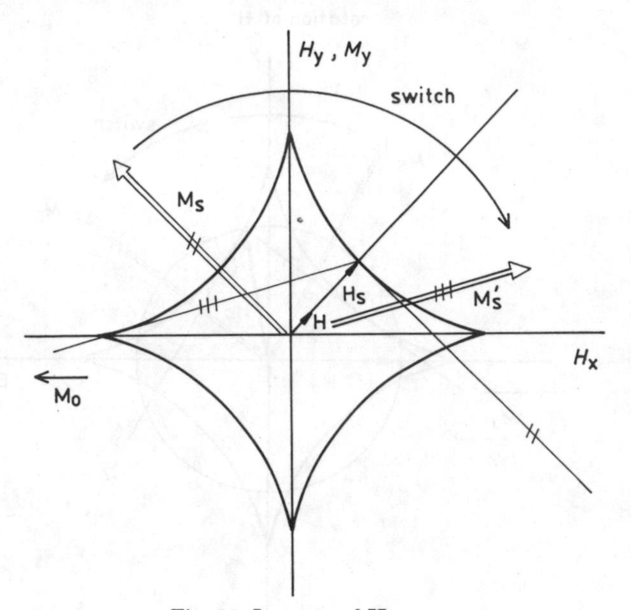


Fig. 86. Increase of **H**.

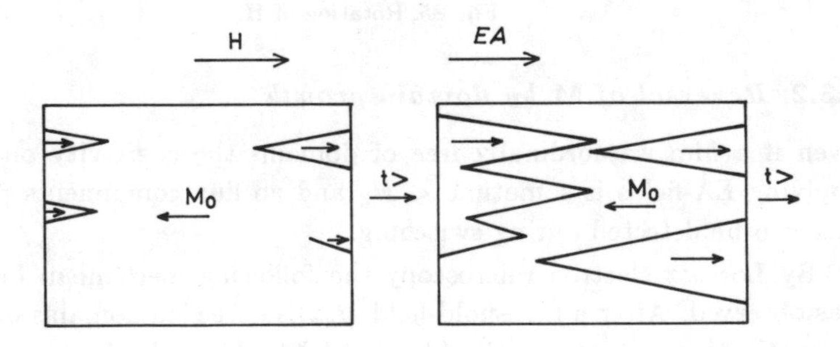


Fig. 87. Domain migration.

So far the quasistatic reversal has been discussed. To sum up the particular features the switching thresholds are plotted in the Stoner Wohlfarth astroid as Fig. 88. One notes:

When $\phi \leq \pi/4$, switching by domain migration occurs.
When $\phi \approx \pi/4$, partial rotation ($H_r$) and completion of reversal by domain migration occur.
When $\phi \leq \pi/2$, coherent-rotation reversal occurs.

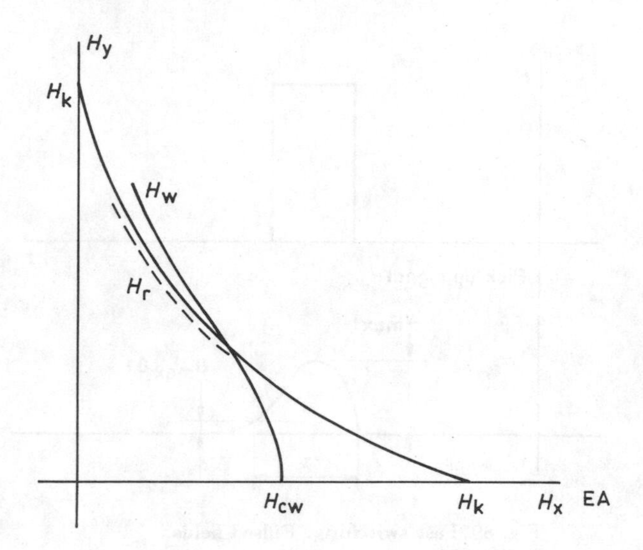


Fig. 88. The particular switching thresholds.

### 8.3.4. *Fast switching* (*pulsed fields*)

The switching time $\tau$ can be studied by cross pick-up coils and high speed film imaging of domains by magneto-optic Kerr effect, as shown in Fig. 89. The data plotted in Fig. 90 are obtained from the following experimental conditions:

Bias field $\perp$ **EA**, $H_y$ = constant.
Pulse field $\|$ **EA**, $H_x$ is a square shape pulse.

Three regions can be distinguished:

I. Reversal by movement of the "diffuse ribbons" (reversal fronts) and by wall motion (walls $\|$ **EA**, wall motion $\perp$ **EA**, diffuse fronts $\|$ **EA**, diffuse front motion $\|$ **EA**).
II. Only diffuse front movement in **EA** direction.
III. Coherent rotation of **M** in entire film.

It is obvious that the coherent rotation is the faster mechanism since in all regions **M** is reversed simultaneously whereas wall motion is a sequential process. The ultimate switching time (mode III) ranging down to nanoseconds is certainly attractive for device fabrication.

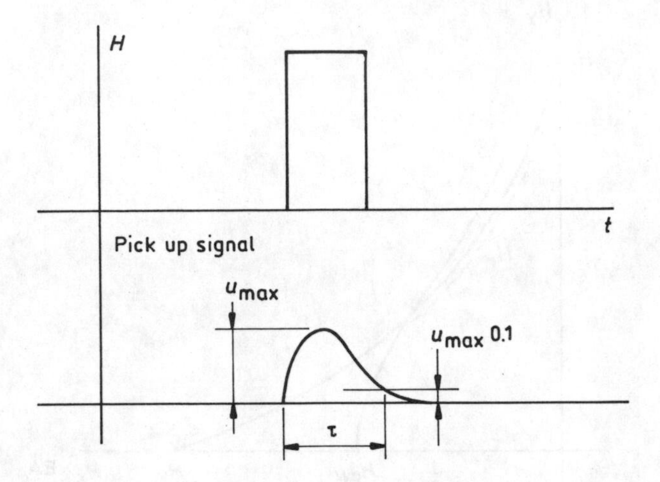


Fig. 89. Fast switching. Pulsed fields.

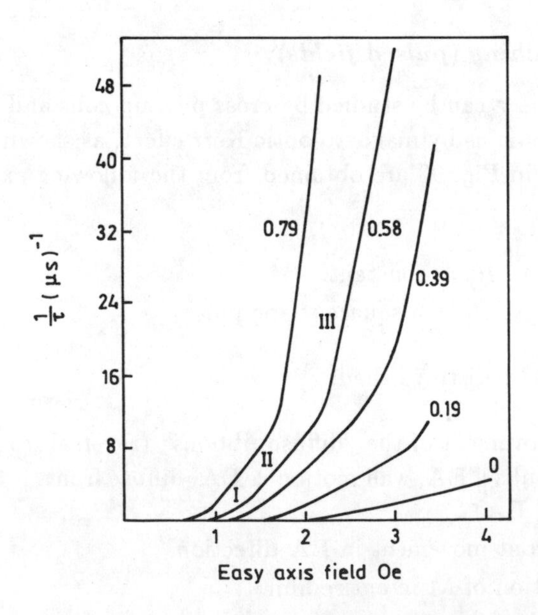


Fig. 90. Three regions of switching speed. Reciprocal of switching time $1/\tau$ vs. EA field.

## 8.4. Applications

### 8.4.1. *Magnetic switching devices*

#### 8.4.1.1. *Preparation*

Permalloy 81Ni19Fe is extensively used. The concentration gives the minimum magnetostriction. Films are produced by evaporation with single boat source and raw material almost of the same composition since vapor pressures differ only slightly. Sputtering is used as well with elevated deposition temperature, perpendicular incidence and magnetic field during deposition. For non-flat films on wires electroplating is used.

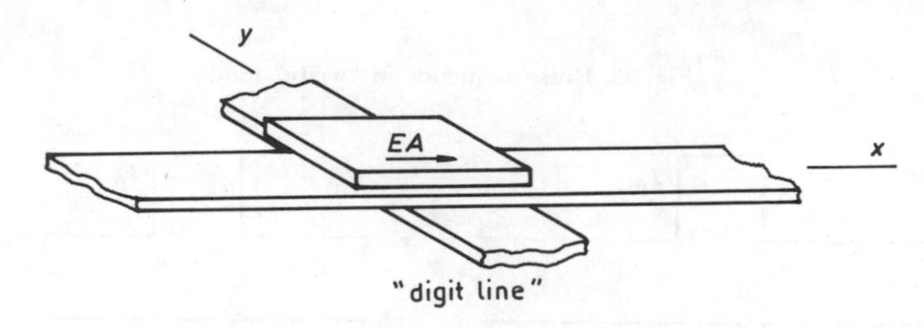


Fig. 91. High speed RAM.

#### 8.4.1.2. *High speed random access memories* (*RAM*)

A single unit is schematically shown in Fig. 91. $H_x$ and $H_y$ are generated by word current $I_w$ (word line) and digit current $I_D$ respectively. The pulse sequence in "write" mode is shown in Fig. 92. $H_w$ twists **M** in **HA** while $H_D$ tips $M$ in +**EA** (store 1) or −**EA** (store 0). In "read" mode the digit line is used for picking up the voltage induced by rotation of **M** as shown in Fig. 93. Reading destroys information since after the rotation of **M** in HA about the same amount of regions rotate clock- or anticlock-wise so that **M** splits into multi-domain state. Combination of RAM units is shown in Fig. 94. The operation range is such that switching must not occur except for both pulses $I_w$ and $I_D$ being present. In the ideal case $I_{D\,min} \geq 0$ ($I_w = \max$) and $I_{D\,max} \leq H_k 2b$ ($b$ being the width of conduction lines) ($I_w = 0$). But in practice $I_{D\,min}$ must switch regions of ripple which are not well aligned in **HA** and $I_{D\,min} \leq H_{cw} 2b$ to avoid the switching by wall motion. $I_w$ usually $> H_k 2b$.

If $\mathbf{H}_w$ is not perpendicular to **EA** the practical margins are as follows. $H_w$ inclines by some degrees against +**EA** after "read 1" is stored. Starting

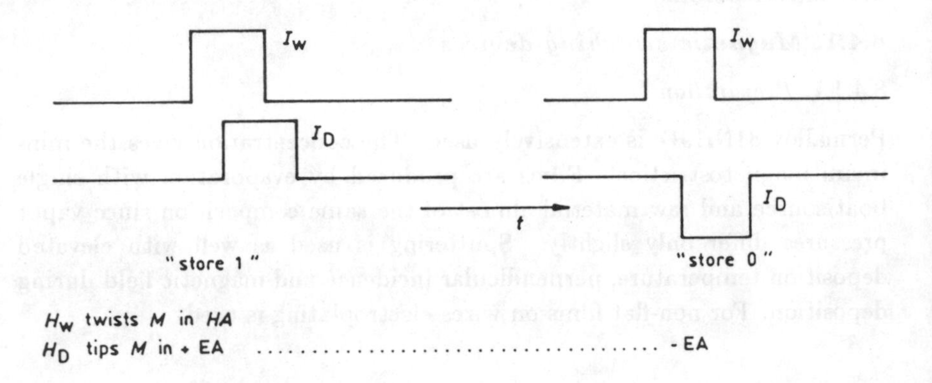


Fig. 92. Pulse sequence in "write" mode.

Fig. 93. Pulse sequence in "read" mode.

from this situation "write 0" is performed with increasing $I_D$ and the signal "read" $U_D$ (Fig. 95) is observed on curve A in Fig. 96. When $I_D$ is reversed at first about 1/2 of ripple domains, then successively more ripple domains, and at last all domains turn in "0". The theoretical upper limit of $I_D$ is the smaller of $H_k 2b$ and $H_{cw} 2b$; however "wall creep" may successively introduce an amount of reversed domains even if $I_w = 0$! $I_{D\,max}$ given by wall creep (curve B in Fig. 96) is shown in Fig. 96. Curve B forms according to the pulse trains shown in Fig. 97. The result is that $I_{D\,min} < I_D < I_{D\,max}$ to ensure that "store A" is really possible but the unwanted "store B" by parasitic digit pulses is avoided. Wall creep can be suppressed by magnetic multilayers or by flux closure using a second magnetic film underneath the strip lines. Non-destructive read-out can be achieved by a similar set-up. One film is of a harder magnetic material, so that in only one film, M is rotated sufficiently by $I_w$. After "read" the hard film aligns M again in the stored direction.

There are some technological limitations. For example, the read-out signal is reduced due to the mutual interaction of single units. The packing density is limited to about 5000 bits/cm$^2$.

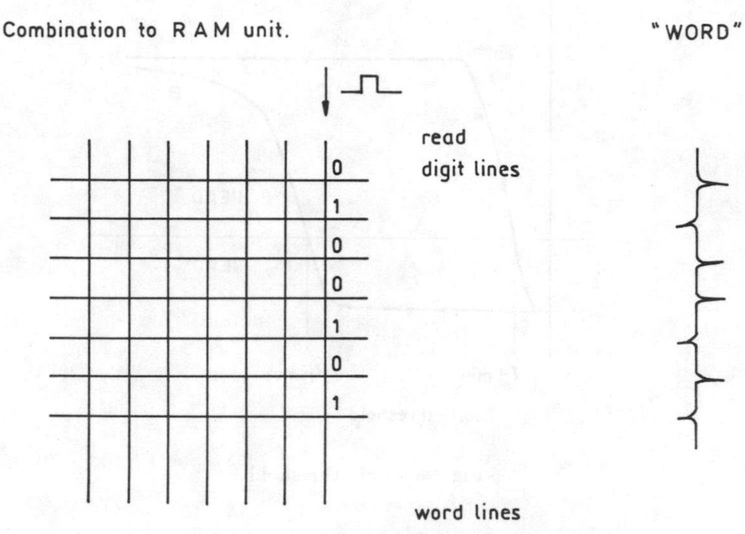


Fig. 94. Combination to RAM unit.

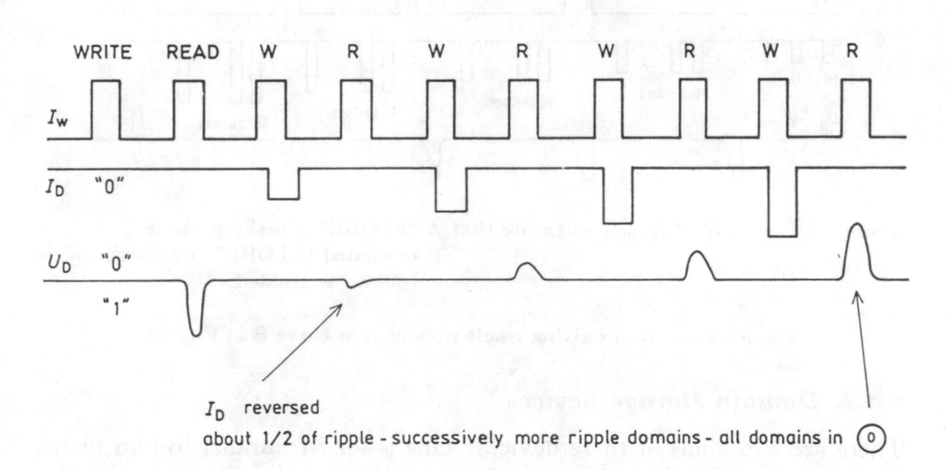


Fig. 95. Pulse trains giving result presented as curve A in Fig. 96.

Plated wire memories give good results. Permalloy 80/20 films 1 micrometer thick electroplated onto BeCu-wires in diameter 0.1 mm are advantageous, having closed flux and high packing densities.

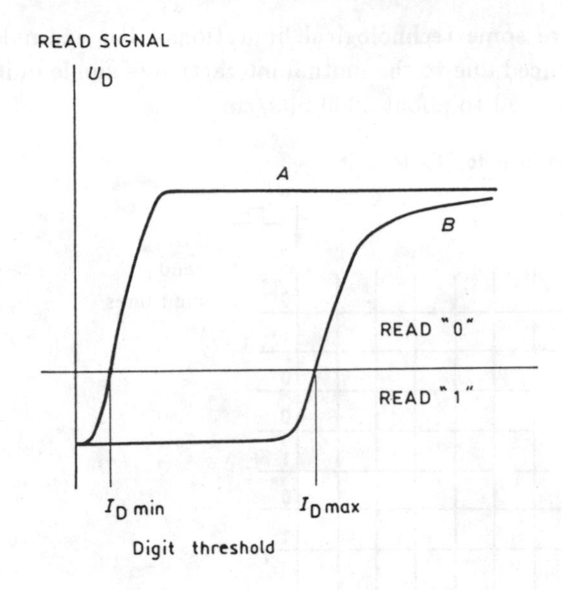


Fig. 96. Digit threshold.

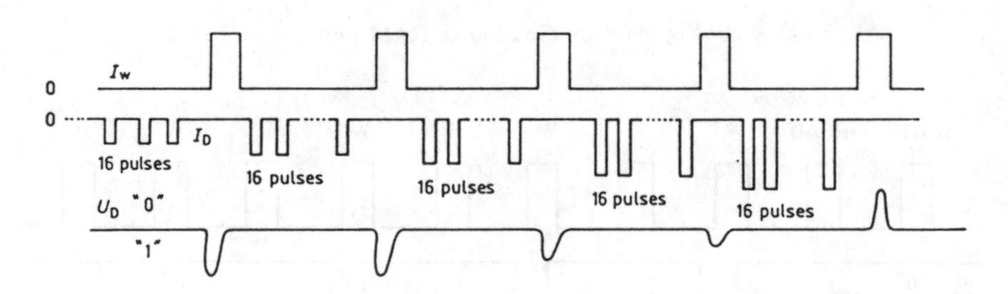


Result : $I_{D\ min} < I_D < I_{D\ max}$ to ensure that A "STORE" is really possible
B unwnated "STORE" (by parasiti digit pulses) is avoided.

Fig. 97. Pulse trains giving result presented as curve B in Fig. 96.

## 8.4.2. *Domain storage devices*

There are two kinds of these devices. One is for **M** parallel to film plane, called domain wall motion devices. The other is for **M** perpendicular to film plane, called "bubble" devices. In both cases domains are moved across the film to the desired position by appropriate fields. This can be done fully electronically, i.e., no moving parts as in drum or disc-memories etc. Bubble devices allow for a high storage capacity and are of great importance.

8.4.2.1. *Bubble domains*

For some special materials thin films with a spontaneous magnetization perpendicular to film surface can be produced, e.g., RE-iron garnets, RE ferrites and amorphous films. The conditions are high anisotropy, constant $K$ and low magnetization $M_s$, since

$$E = K \sin^2\theta + 2\pi M_s^2 \cos^2\theta$$

$E = E_{\text{min}}$ when $\partial E/\partial\theta = 0$ and $\partial^2 E/\partial\theta^2 > 0$, for which $\theta = 0$ or $\pi$ yields

$$K > 2\pi M_s^2 \qquad \text{or} \qquad H_k > 4\pi M_s \ .$$

Fig. 98 illustrates the orientations.

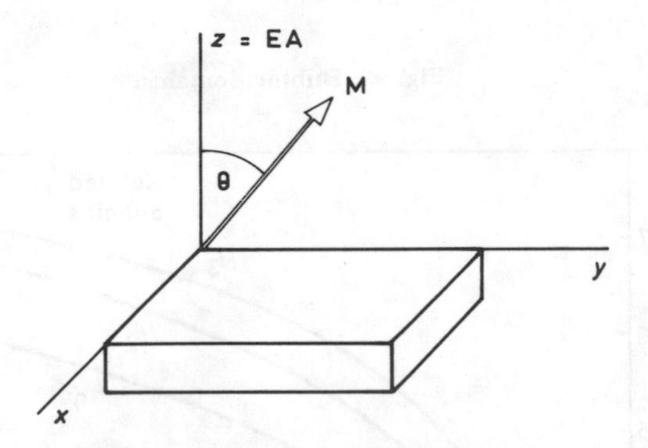


Fig. 98. Magnetization for bubble domains.

Stray field minimization leads to strip domains of antiparallel orientation. By applying external fields $\mathbf{H} = H_z e_z$, unfavorable oriented strips shrink, pinch off and finally the spherical domains become "bubbles", as shown in Fig. 99. The determination of equilibrium shape and distribution of domains is a complex variation problem in which a characteristic length

$$l = E_{\text{wall}}/4\pi M_s^2$$

appears.

The calculation leads to the plot of field margins limiting special domain structures for differently thick films. As $\mathbf{H}$ rises, strip lattice changes to a bubble lattice at $H_1$. When bubble spacing is large enough to inhibit interaction, "free bubbles" are formed at $H_2$ which are stable up to $H_3$

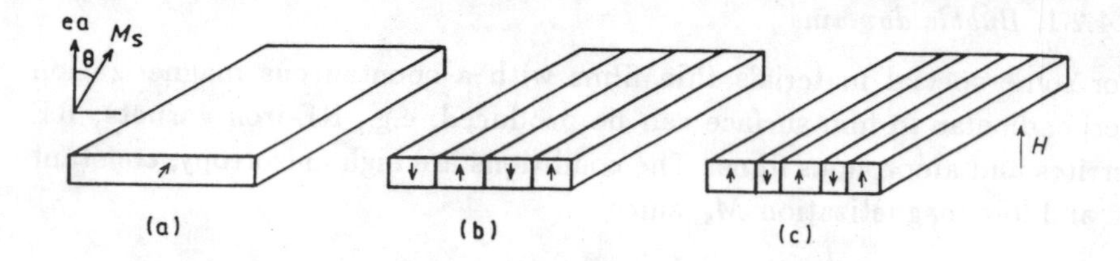


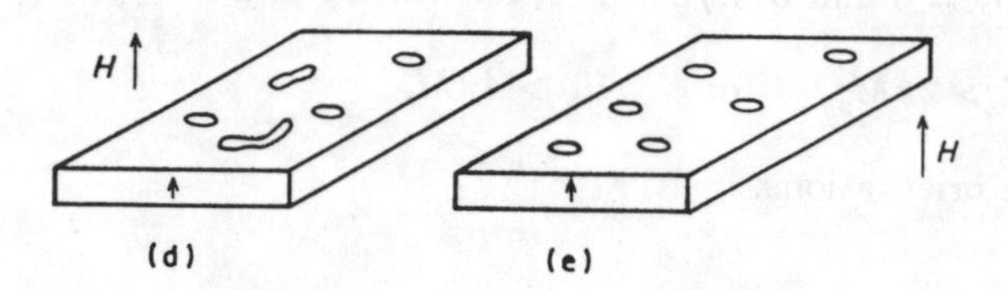


Fig. 99. Bubble domains.

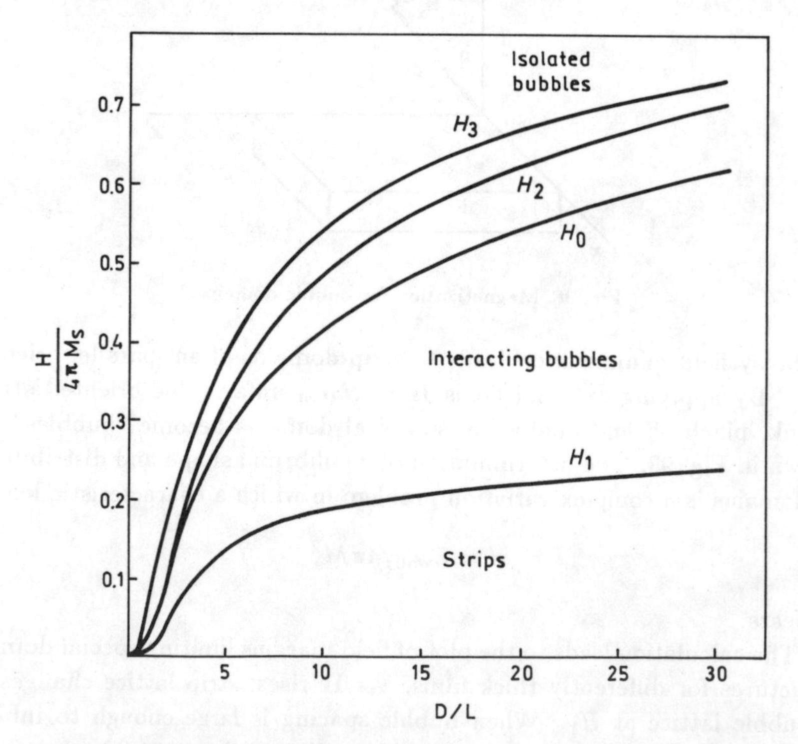


Fig. 100. Field margins limiting special bubble domain structures for differently thick films

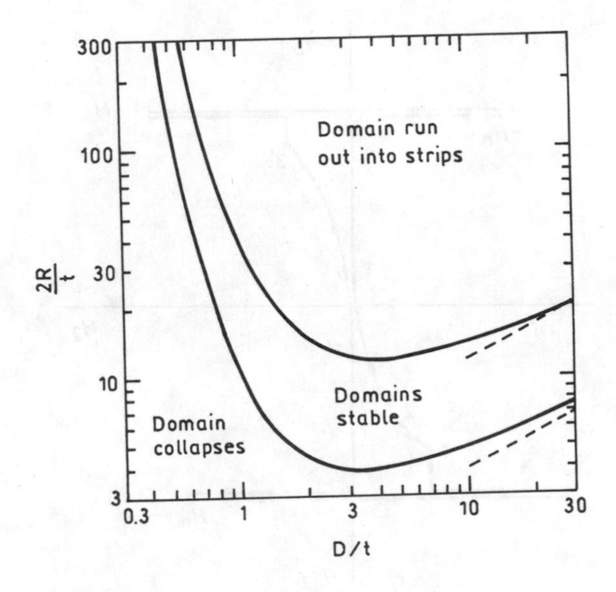


Fig. 101. Reduced bubble diameter $2R$ at the collapse and runout fields as a function of reduced film thickness.

above which they collapse. When the field is reduced free bubbles run out to strips again at $H_0$. The bubble diameter is of technical importance for storage capacity (dense packing). It is plotted also as a function of thickness for the field limits $H_0$ and $H_3$; see Fig. 100 and 101. $D$ is the film thickness. Optimal values are obtained for $D = 4l$ where the bubble radius is $R = 4l$. Bubble boundaries are Bloch walls of either unique or a split sense of rotation along the circumference. The latter leads to "hard bubbles" which require higher fields for their motion. Figure 102 shows the hysteresis loop of bubble materials. Starting from saturation $M_z > 0$ by $H_z > 0$ we decrease $H$ to 0. $M$ remains saturated metastable to $H = -H_n$ but if even locally the threshold $H_n$ for wall nucleation is exceeded the entire film demagnetizes by the formation at strip domains (1 in Fig. 102). Reducing $H_z$ further, bubbles are formed which collapse at 2 in Fig. 102 and the film will saturate in the reverse direction. Various materials and films are shown in Fig. 103. Orthoferrites $REFeO_3$, hexagonal ferrites $MeFe_{12}O_{19}$, and garnet $REFe_5O_{12}$ form epitaxial films of a suitable

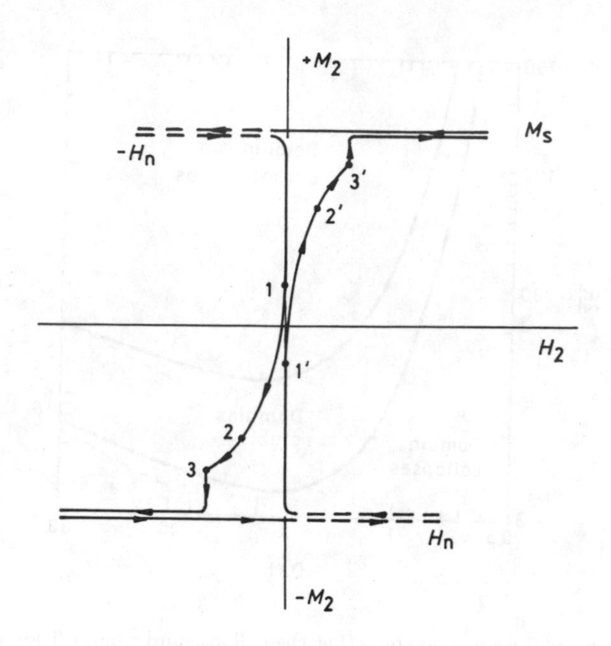


Fig. 102. Hysteresis loop of bubble materials.

bubble size and mobility by liquid phase epitaxy and CVD. Amorphous films GdCo, NdFe, GdFe etc. with bubbles of diameters in micrometers can be obtained. $M_s$ is fairly high needing high bias fields, which can be achieved by making ternary alloys GdCo (Au, Cu or Mo) show better properties, approaching those of garnets.

Bubble propagation is realized by means of two mechanisms. One is by local fields produced by thin film conductor loops or meanders. The other is by rotating fields magnetizing magnetic film bars. (e.g., permalloy). Both additional structures are deposited onto the bubble films where bubbles follow the fields like a freely movable magnet. See Fig. 104. High capacity storage devices are invariably of the major-minor loop variety. In such a unit bubbles are introduced locally by a bubble generator and at a transfer gate into the major loop when required. Transfer into the minor loops takes place at other gates. Read-out is achieved either by magnetic resistance, Hall effect or magneto-optically. The first is most promising since permalloy is used which is also necessary for the magnetic bar structure. Usually the bit rates are in the MHz region and the packing densities are around $10^6$ bits/cm$^2$. See Fig. 105.

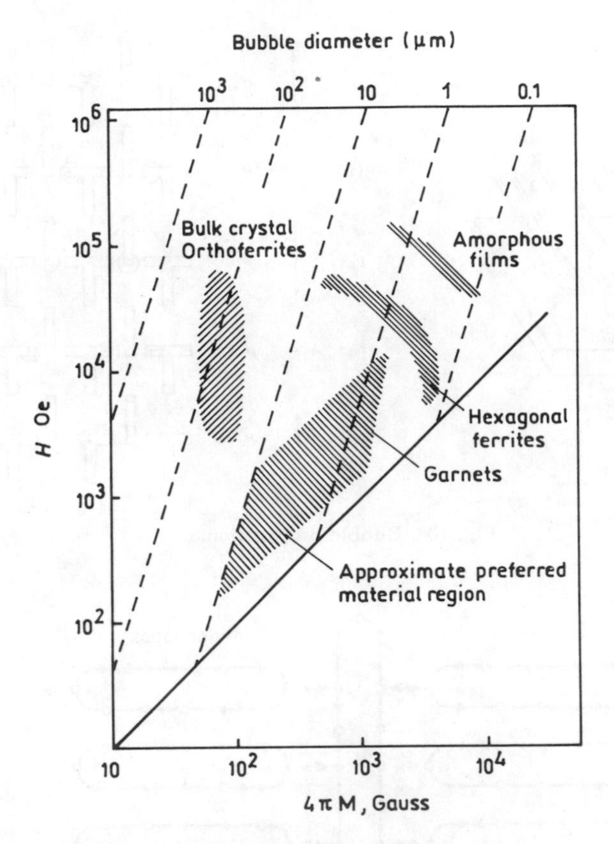


Fig. 103. Various bubble materials.

8.4.2.2. *Domain wall motion devices*

Thin uniaxial films, **EA** in the plane, are overlaid by special conductor structures. Domains of reverse direction can be shifted by special pulsed currents in both metallic meanders deposited onto the magnetic film. The fields are to be taken in the range $H_{cw} < H < H_k$, i.e., not to generate new unwanted domains and to move the domain as fast as possible. The limits on performance are in fact the spurious domains which may occur at the edges or at nuclei left by the trailing edges of the moving domains. Other solutions using specially alternating fields generated by current lines are possible. A sawtooth pattern of low coercivity NiFeCo in a film of high coercivity is a good choice, as seen in Fig. 107. Such structures are obtained if the complementary sawtooth pattern is first deposited in the form of rough Al film onto glass substrates, which is then overlaid

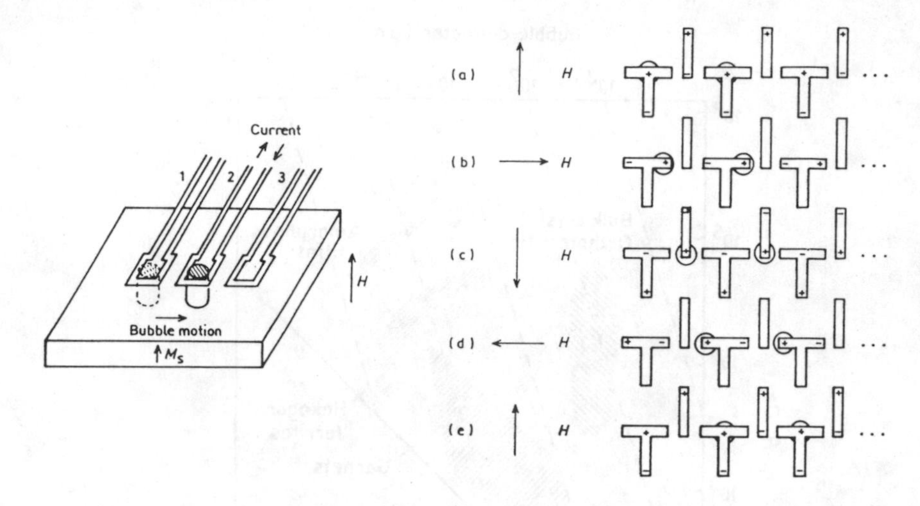


Fig. 104. Bubble propagation.

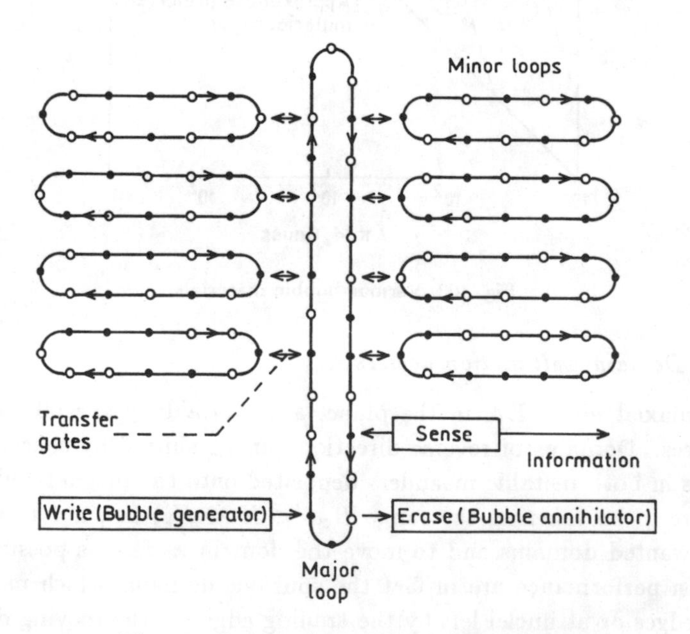


Fig. 105. A major/minor loop bubble store.

by the magnetic film. The bit rates of such domain shift registers are 500 kHz and bit densities are 300 bits/cm$^2$. 24M bit memories have been constructed. Finally the optical beam storage devices should be mentioned at least regarding its principle. $EA \perp$ film plane (e.g., MnBi). Laser pulses heat the film locally above the Curie point. Then the flux closure through this paramagnetic part causes the magnetization to reverse during cooling down again. "Curie point writing" erasure is performed by high reversing fields (700 Oe) which cannot be applied as locally as to erase specific spots. Therefore smaller erasure fields plus laser pulses are suitable. Films are deposited on discs and drums. Other materials are EuO, CrTe, GdCo, MnAs and others, see Fig. 108.

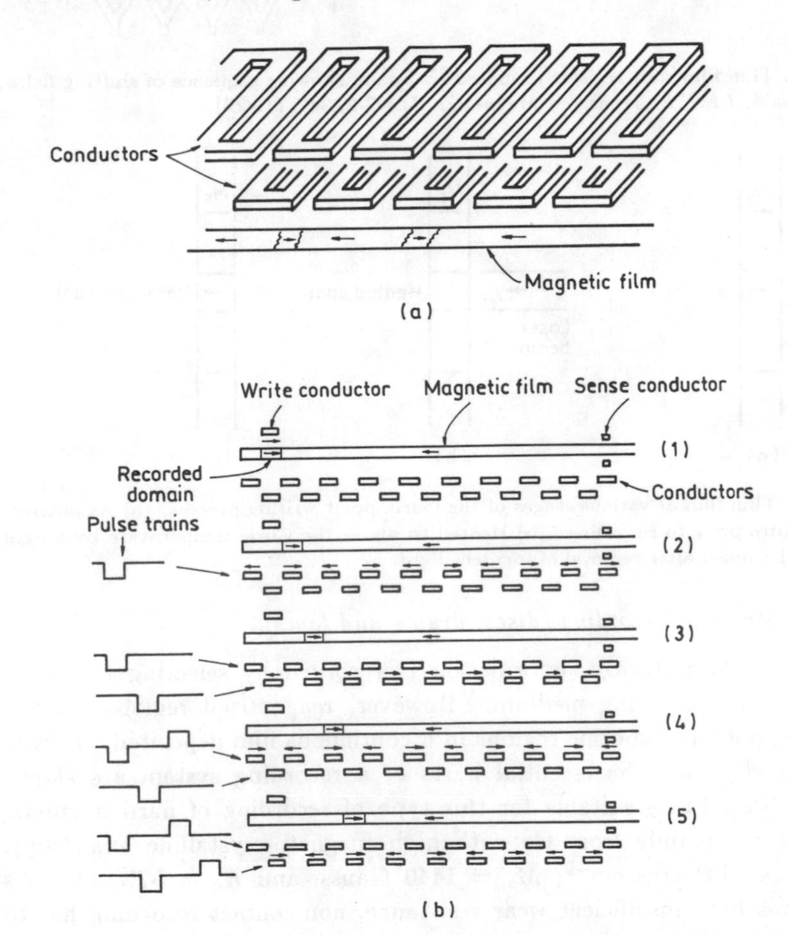


**Fig. 106. Thin film shift register proposed by Broadbent. (a) Exploded view of conductor and film pattern. (b) Cross section view of conductors and film showing field and magnetization directions.**

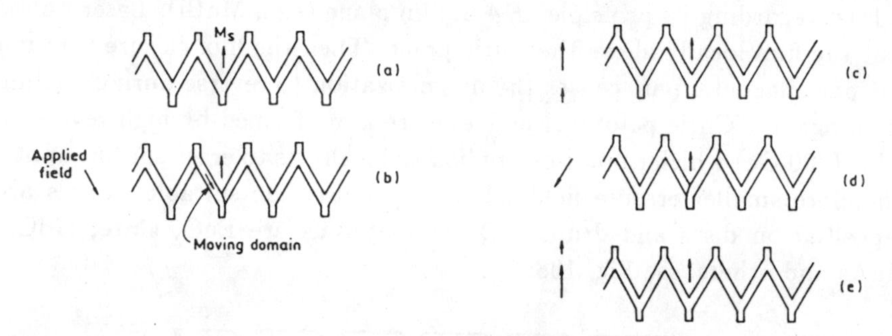


Fig. 107. Thin film shift register proposed by Spain, showing sequence of shifting fields. [R. J. Spain, *I.E.E.E. Trans. on Magnetics, MAG*-**2**, 347, (1966)]

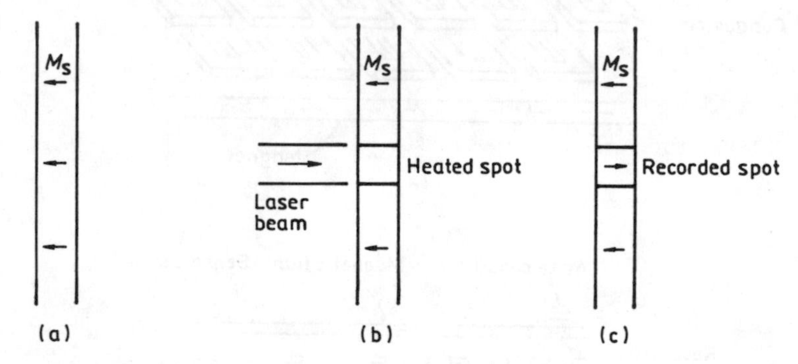


Fig. 108. Thin film at various stages of the Curie point writing process. (a) At ambient temperature prior to recording. (b) Heated to above the Curie temperature by a light pulse. (c) Cooled after removal of incident light.

### 8.4.2.3. *Magnetic recording (discs, drums and tapes)*

As in thin film storage, information is recorded by selecting the sense of **M** in the recording medium. However, magnetized regions are not discrete spots, but specific regions in a continuous film deposited on discs, tapes or drums. The essential parts of a recording system are shown in Fig. 109. Films suitable for this type of recording of hard magnetic material are mainly from Co with high magneto-crystalline anisotropy. $K_{111} = 4 \cdot 10^6$ ergs cm$^{-2}$, $M_s = 1440$ Gauss, and $H_c = 5700$ Oe. As such films have insufficient wear resistance, non-contact recording has to be applied. This reduces the information density. Recording material has been evaporated under oblique incidence at an elevated temperature which leads to a microstructure exhibiting arrays of columns. Thus a high particle-shape anisotropy is formed.

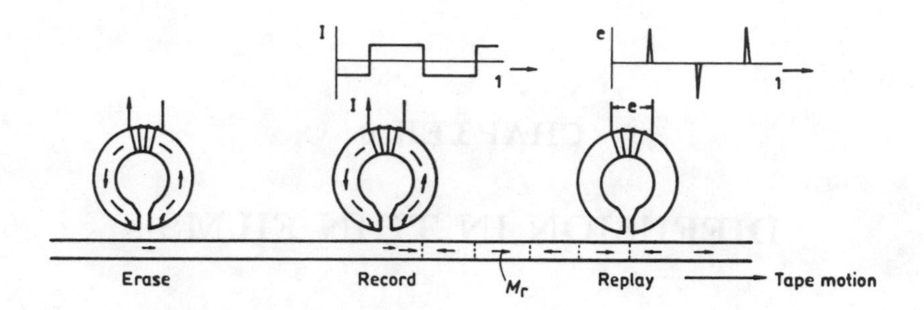


**Fig. 109. Simplified diagram of a digital recording system.**

Other methods of film deposition are electroless chemical plating from Co-chloride, Na-hypophosphite, citric acid, and Na-laurel-sulphate, and electrodeposition in similar baths.

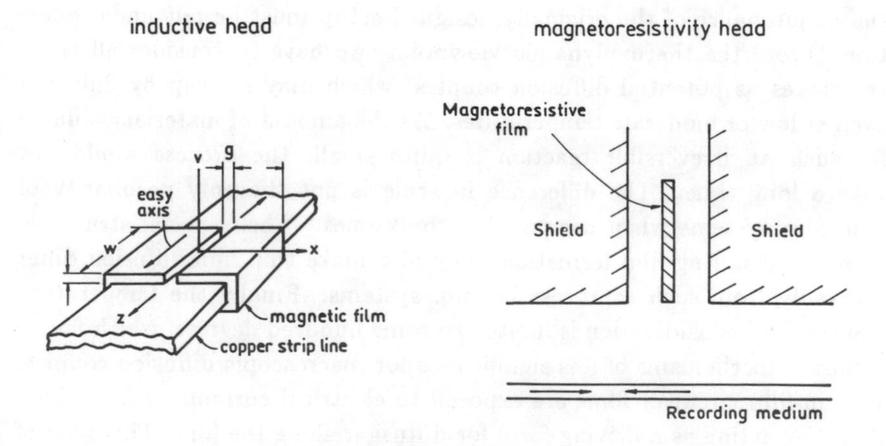


**Fig. 110. Other devices made of thin magnetic films.**

Not only the recording media but also the record, read and erase heads can be produced in thin film form, as shown in Fig. 110.

# CHAPTER 9

# DIFFUSION IN THIN FILMS

## 9.1. General

Whenever thin films are involved in devices or are simply used as coatings the maintenance of the originally designed setup must be put under question. From the thermodynamic viewpoint, we have to consider all these structures as potential diffusion couples, which may mix up by diffusion even at low or moderate temperatures. As the amount of material available for such an irreversible reaction is quite small, the process would not take a long time. The difference in scale is not the only peculiarity of thin film systems when compared to bulk ones. Their special structure, introduced during film formation, may also make thin film diffusion differ markedly from such a process in bulk systems. Finally the temperature range under consideration is limited to some hundred degrees, which favors diffusion mechanisms of less significance for macroscopic diffusion couples. In thin-film circuitry films are exposed to electrical currents of fairly high densities acting as a driving force for diffusion along the film. This type of material transport known as electromigration is often the cause of failure in integrated circuits. Hence, for reliability and lifetime reasons thin film diffusion plays an important technological role.

On the other hand, diffusion may become a useful tool to form alloys or compounds by reaction of films or of a film and its substrate. An example is the formation of silicides (PtSi) sandwiched between the semiconductor and the metallization to ensure reliable and low resistance contacts. This inserted layer formed by diffusion of a Pt deposit on Si acts as a barrier against further reaction between Si and the metallization (mostly Al).

## 9.2. Fundamental

The basic principles of atomic migration considered either as a sequence of discrete process or on a large scale on the basis of continuous concentration variations are first discussed on the basis of homogeneous crystal lattices. Further on, we shall include discontinuities like surface and grain boundaries which play an important role in thin film diffusion.

### 9.2.1. *Energy considerations*

#### 9.2.1.1. *Activation energy*

Diffusion in a crystal lattice implies the presence of point defects such as interstitials or vacancies. Their formation requires energy but their existence is associated with an increase of entropy by disorder, so that the probability of their occurrence is larger than zero. The defect density at temperature $T$ is given by

$$n = \exp(-g_\mathrm{f}/kT) = \exp[(s_\mathrm{f}T - E_\mathrm{f})/kT]n_0 \exp(-E_\mathrm{f}/kT)$$

where $g_\mathrm{f}$ is the Gibbs free energy of formation, $s_\mathrm{f}$ the entropy of formation, and $E_\mathrm{f}$ the measurable energy of formation. For migration of the defect a barrier of height $E_\mathrm{m}$ has to be surmounted. This is attempted $\omega_0$ times per unit time, hence the jump frequency is given by

$$\omega = \omega_0 \exp(-E_\mathrm{m}/kT)$$

for the defect. Now, in the case of a vacancy mechanism for an atomic jump the frequency is limited by the probability to find a neighboring vacancy, hence

$$\nu = \omega n = n_0\omega_0 \exp[-(E_\mathrm{f} + E_\mathrm{m})]/kT = \nu_0 \exp(-E_\mathrm{A}/kT) \,.$$

The same expression holds for an interstitial mechanism. For diffusing interstitials the energy of formation has to be provided only once. $E_\mathrm{A}$ is called the activation energy.

#### 9.2.1.2. *Energy barriers and jump frequencies in a driving force*

The equivalence of both directions for atomic jumps is disturbed by a driving force as seen in Fig. 111(b). The slope of the basic line of the

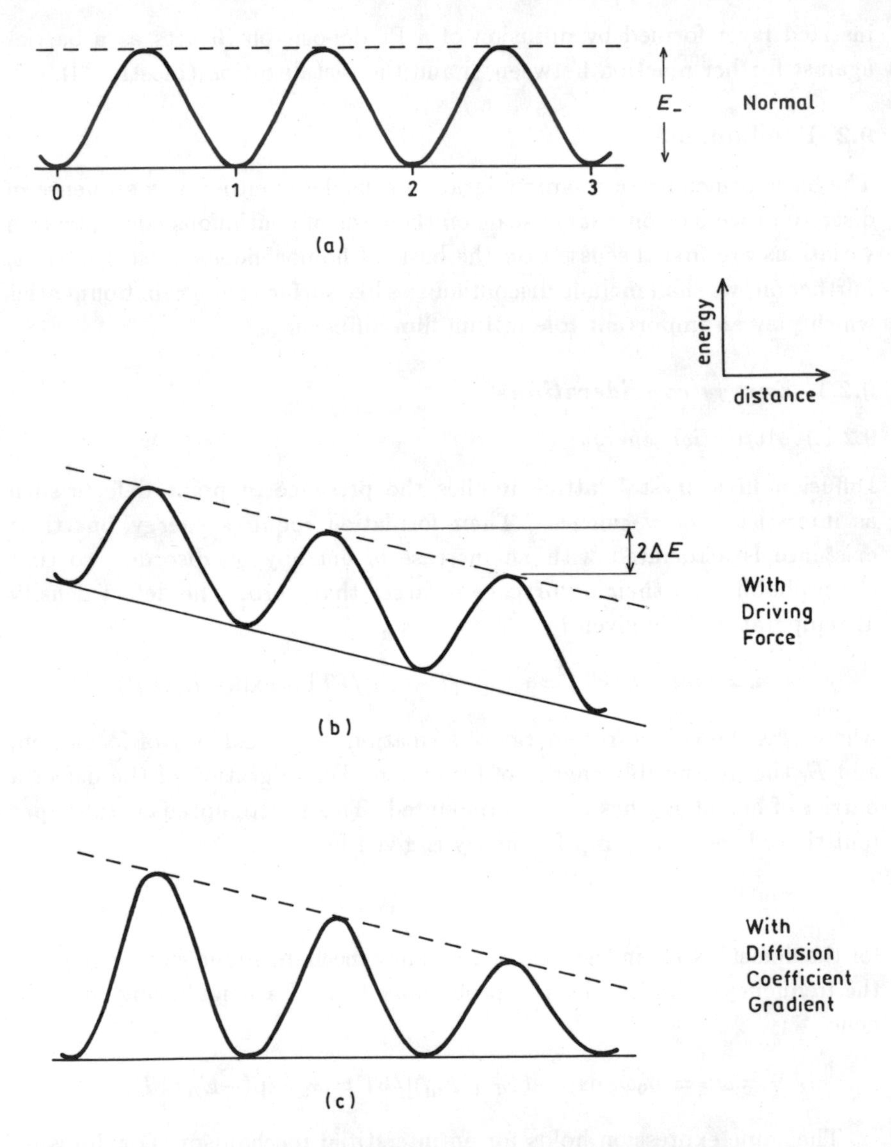


**Fig. 111. Atomic potential energy versus distance.**

periodic potential energy is the driving force $F$, decreasing the energy of jumps in positive direction by $\Delta E_{\rm m} = 1/2\lambda F$, where $\lambda$ is the jump distance. Vice versa, the barrier is increased correspondingly for jumps in the opposite direction. The jump frequency in the force direction is

$$\Gamma_+ = \Gamma_0 \exp(\Delta E_m / kT)$$

where $\Gamma_0$ is the jump frequency in the absence of driving force. Normally $\Delta E_m \ll kT$, so

$$\Gamma_+ \approx \Gamma_0(1 + \lambda F/2kT)$$

$$\Gamma_- \approx \Gamma_0(1 - \lambda F/2kT)$$

where $\Gamma_-$ is the frequency in the opposite direction.

The mean migration distance after $N$ jumps is given by

$$\langle x \rangle = Np\lambda - Nq\lambda$$

where $p$ and $q$ are the probabilities for positive and opposite jumps

$$p = \Gamma_+/(\Gamma_+ + \Gamma_-) = \Gamma_+/2\Gamma_0, \qquad q = \Gamma_-/2\Gamma_0 .$$

Then

$$\langle x \rangle = N\lambda^2 F/2kT .$$

### 9.2.2. *Average atomic displacements—diffusion law*

Consider the concentration of atoms in a given network of sites at position $\mathbf{x} = (x_1, x_2, x_3)$ at time $t : c(\mathbf{x}, t)$. It is altered with time $\tau$ by atoms moving in and out the region around $\mathbf{x}$. With the probability $W(\mathbf{X}, \tau)$ for an atom to migrate along $\mathbf{X}$ in time $\tau$ we may write

$$c(\mathbf{x}, t + \tau) = \sum_{\text{all } \mathbf{x}} W(\mathbf{x}, \tau) c(\mathbf{x} - \mathbf{X}, t) .$$

A Taylor expansion around $t$ and $\mathbf{x}$ yields

$$c(\mathbf{x}, t) + \tau \frac{\partial c}{\partial t} = c(\mathbf{x}, t) - \left[ \sum_{\text{all } \mathbf{x}} W(\mathbf{x}, \tau)\mathbf{x} \right] \cdot \nabla c$$

$$+ \sum_i^3 \sum_j^3 \frac{1}{2} \frac{\partial^2 c}{\partial x_i \partial x_j} \left[ \sum_{\mathbf{x}} W(\mathbf{X}, \tau) \mathbf{X}_i \mathbf{X}_j \right]$$

$$\tau \dot{c} = -\langle \mathbf{X} \rangle \cdot \nabla c + \frac{1}{2} \sum_i \sum_j \frac{\partial^2 c}{\partial x_i \partial x_j} \langle \mathbf{X}_i \cdot \mathbf{X}_j \rangle$$

$$\dot{c} = -\frac{1}{\tau} \langle \mathbf{X} \rangle \cdot \nabla c + \sum_i \sum_j D_{\text{ij}} \frac{\partial^2 c}{\partial x_i \partial x_j}$$

where

$$D_{ij} = (1/2\tau)\langle \mathbf{X}_i \cdot \mathbf{X}_j \rangle ,$$

is called diffusion tensor. It is always symmetric and its main axes are simply that of the crystal in the case of orthorhombic axes. If these axes are chosen as coordinate axes the main diffusivities are $D_{\mathrm{I}}, D_{\mathrm{II}}, D_{\mathrm{III}}$ which reduce to one value $D$ for cubic crystals. Then in the case of random walk without driving forces or diffusion coefficient gradients, we have $\langle \mathbf{X} \rangle = 0$. So

$$\dot{c} = D\Delta c$$

which is called Fick's second law, where

$$D = \langle X_i^2 \rangle / 2\tau = \langle \mathbf{X}^2 \rangle / 6\tau$$

since

$$\langle \mathbf{X}^2 \rangle = \langle X_1^2 + X_2^2 + X_3^2 \rangle = \langle X_1^2 \rangle + \langle X_2^2 \rangle + \langle X_3^2 \rangle = 3\langle X_i^2 \rangle$$

In the presence of a driving force,

$$\langle \mathbf{X} \rangle \approx \frac{N}{3} \lambda^2 F / 2kT$$

because among $N$ total jumps only one third are along the $\mathbf{F}$ direction. With

$$D = \langle X_i^2 \rangle / 2\tau = \langle X^2 \rangle / 6\tau = \frac{N}{3} \lambda^2 / 2\tau$$

the diffusion law writes

$$\dot{c} = -\frac{D\mathbf{F}}{kT} \cdot \nabla c + D\Delta c = -\langle \mathbf{v} \rangle \cdot \nabla c + D\Delta c$$

or, in terms of fluxes,

$$j = \mathbf{F} Dc / kT - D\Delta c$$

which is valid for $D = D(x)$ as well.

### 9.2.3. *Atomic fluxes and diffusion coefficients*

$j = -D\nabla c$ defines the "diffusion coefficient" regardless of what forces and gradients are present.

Tracer diffusion coefficient $D^*$ is the diffusion coefficient describing the diffusion process where only one species of atom is considered as diffusant.

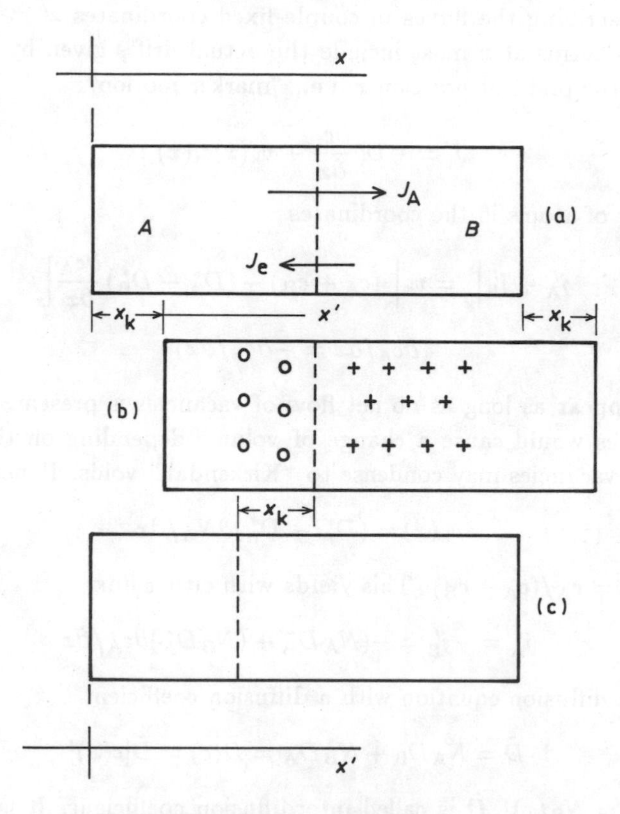


Fig. 112. Interdiffusion. "Marker" motion.

Such a situation is, e.g., given in high dilution of the diffusant in a host matrix.

Interdiffusion is a process where two species of atom intermingle by mutual diffusion of comparable quantities. Two fluxes of atoms must be considered. It should be noticed that viewing thin film diffusion in nearly all cases the diffusion process is essentially of this type! We consider the situation according to Fig. 112. In each cross section,

$$J_i = -D_i^* \nabla c_i$$

applies. If $D_A^* \neq D_B^*$, e.g., $D_A^* > D_B^*$, then everywhere we have $j_A > j_B$ and more material is transported to the left, causing a drift $v_k(x)$ of the whole system if the couple is fixed in the coordinate system at plane $x$ which need not necessarily be the original interface. This drift is called "Kirkendall drift". Describing the fluxes in couple-fixed coordinates $x' = x - x_k$ the actual flow terms at $x$ must include the actual drifts given by the reverse motion of the plane at position $x$, i.e., "marker motion":

$$J_i' = -D_i^* \frac{\partial c_i}{\partial x} + v_k(x) c_i(x) .$$

A net flow of atoms in the coordinates

$$j_A' + j_B' \Big|_x = v_k \Big|_x (c_A + c_B) - (D_A^* - D_B^*) \frac{\partial c_A}{\partial x} \Big|_x$$

$$(\partial c_A / \partial x = -\partial c_B / \partial x)$$

must disappear as long as no net flow of vacancies is present. A net flow of vacancies would cause a change of volume depending on the position. Excessive vacancies may condense to "Kirkendall" voids. Hence

$$v_k(x) = (D_A^* - D_B^*) \partial N_A / \partial x$$

where $N_A = c_A/(c_A + c_B)$. This yields with either flux

$$j_A' = -j_B' = -(N_A D_B^* + (N_B D_A^*) \partial c_A / \partial x$$

which is a diffusion equation with a diffusion coefficient

$$\tilde{D} = N_A D_B + N_B D_A = \tilde{D}(c) = \tilde{D}[c(x)]$$

since $N_A = N_B(x)$. $\tilde{D}$ is called interdiffusion coefficient. It is reduced to $\tilde{D}$ = constant only if $D_A = D_B$. In principle one can deduce $\tilde{D}(c)$ from the measured concentration profiles if the Boltzmann-Matano transformation can be applied. This is possible for thin films only as an approximation. Since lattice defects give rise to a rather inhomogeneous spatial distribution of the concentration it is reasonable to ignore the concentration dependence of $D$ and replace it by an average "lattice diffusivity $D_l$", or even by an "effective interdiffusivity" also including grain boundary diffusion coefficients in the average. The latter value then describes only the flux across the original couple interface but does not account for further atomic distributions. It is, however, a convenient figure to give quick estimate of mixing rates. For single crystal film diffusion, marker motion can be observed to determine the differences in diffusion speed across the formed compound layers, see Fig. 113.

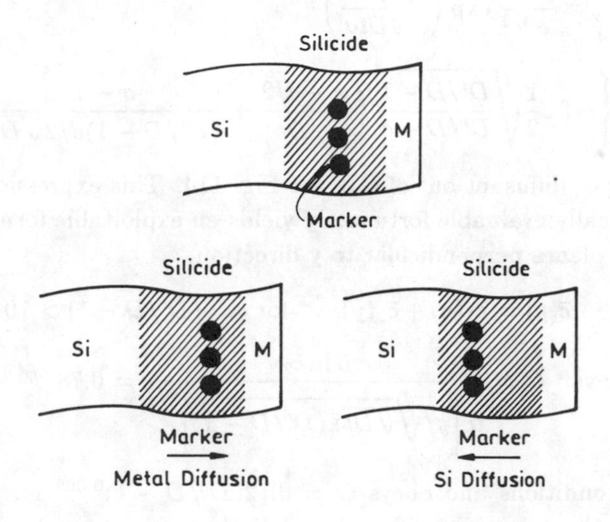


**Fig. 113. Marker motion in silicide formation.**

## 9.3. Diffusion in Structural Inhomogeneous Systems

Particularly in thin film growth, induced lattice imperfections are present to an extent overthrowing bulk solids by far. On the other hand very often the growth mechanisms lead to more or less geometrically well defined defect arrangements which favors analytical or at least numerical treatment of diffusion in such systems. In the following we shall briefly discuss the effect of loosely packed regions such as interfaces and grain boundaries on interdiffusion in thin films, both of extraordinary importance for thin film diffusion properties.

### 9.3.1. *Analytical approaches to grain boundary (GB) diffusion*

Whipple as well as Suzuoka gave an analytical treatment of the superposition of lattice and GB diffusion using Fourier-Laplace transformation. The problem was stated for an infinite half-space split by a GB plane. The solution outside the proximity of the GB is given by the superposition of two concentrations $c = c_1 + c_2$, where

$$c_1 = c_0 \operatorname{erf}(y/\sqrt{Dt})$$

due to material transfer across the interface, and $c_2 = \frac{yc_0}{\sqrt{4\pi Dt}}$ INT, where

$$\mathrm{INT} = \int_1^{D'/D} \frac{d\sigma}{\sigma^{3/2}} \exp\left(-\frac{y^2}{4Dt\sigma}\right)$$

$$\times \left\{ \operatorname{erf} c \frac{1}{2}\sqrt{\frac{D'/D-1}{D'/D-\sigma}} \left(\frac{x-d/2}{\sqrt{Dt}} + \frac{\sigma-1}{(D'/D-1)d/2\sqrt{Dt}}\right)\right\}$$

due to leaking of diffusant out of GB, see Fig. 114. This expression which is only numerically evaluable fortunately yields an exploitable form when $c$ is averaged in planes perpendicular to y direction,

$$\bar{c}(y) = \bar{c}_1(y) + \bar{c}_2(y) \qquad \text{for} \qquad (D'/D - 1) > 10 \, .$$

$$\bar{c}_2 \quad \text{obeys} \quad G = \frac{\partial \ln \bar{c}_2}{\partial \left[ y/\sqrt{\sqrt{Dt}\frac{d}{2}(D'/D-1)} \right]^{6/5}} = 0.78$$

for Whipple conditions and obeys $G = 0.72(D'/D - 1)^{0.008}$ for Suzuoka conditions. Both expressions differ practically by a few percent, indicating that the problem is not sensitive to boundary conditions. If we neglect $c_1$ outside a range $y \geq \sqrt{Dt}$, then from the slope of the depth profile given in terms of

$$\ln \bar{c} \qquad \text{versus} \qquad y^{5/6}$$

one obtains

$$\left(\frac{D'}{D} - 1\right) \frac{d}{2\sqrt{Dt}}$$

using the above equations. It finally gives $D_b$. For GB width usually 0.5 nm is assumed by convention. $D$ may be evaluated from the slope of $\bar{c}(y)$ at the interface where $c_2$ may be ignored.

## 9.4. Application to Thin Films[16–19]

### 9.4.1. *Whipple treatment of thin film depth profiles*

For thin films this treatment is not directly suitable since limited thickness dams up the diffusant in GB. Furthermore an array of GB gives different values according to the distance in relation to the lateral diffusion paths. Such features are accounted for in refined treatments which show that errors of the order of the magnitude of $D$ are introduced if Whipple's conditions are drastically unfulfilled. Hence one should apply the analysis for fairly

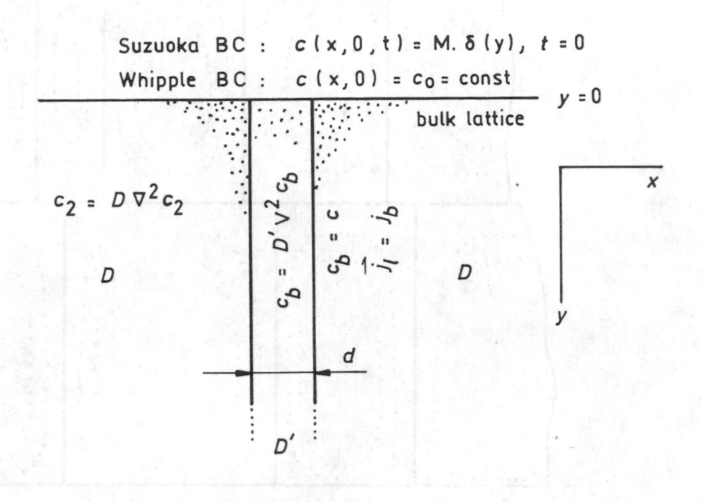


Fig. 114. Grain boundary diffusion.

short diffusion time, when the diffusant has not yet significantly penetrated through GB:

$$\left(\frac{D'}{D} - 1\right) \frac{d\sqrt{Dt}}{2} \leq 0.1l^2 \qquad (l = \text{ layer thickness}).$$

The isoconcentration contours in the case of many GB's are shown in Fig. 115.

Besides Whipple's treatment of depth profiles, GB diffusivities in thin films may be calculated also for very early diffusion states from the amount of diffusant dispersed at the opposite surface after, crossing via GB surface where the concentration is $\bar{c}$. However, in that case the GB density must be known. With the spacing $L$ of GB's normalized by $\sqrt{Dt}$ one can deduce $(D'/D - 1)\frac{d\sqrt{Dt}}{2l^2}$ and then $D'$ from the Gilmer–Farrell curve which holds for large ratios $D'/D$, usually larger than one thousand, see Fig. 116.

### 9.4.2. *Short circuit diffusion — analytical approach to GB diffusion in thin films*

In late diffusion states, the information about $D'$ has almost completely disappeared in the concentration distribution. The concentration gradients along GB's and surfaces are smoothed out. These diffusion paths are nearly ideal short circuits. The rate-limiting factor is only the lattice diffusivity;

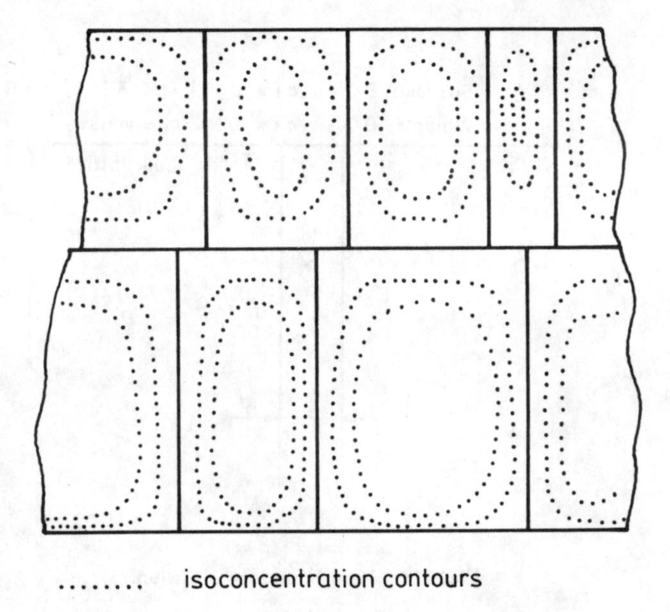


Fig. 115. Isoconcentration contours.

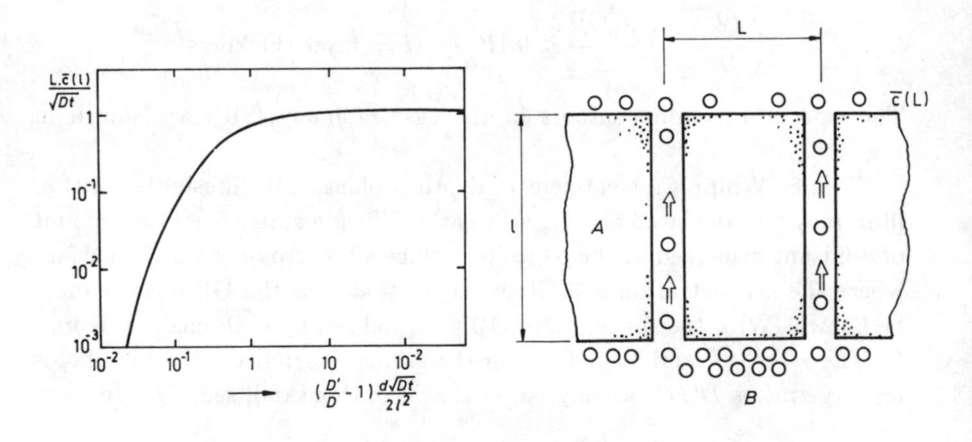


Fig. 116. Gilmer–Farrell curve.

however the diffusion geometry is not simply a planar one. Hence, for orthoprismatic grains on whose surface $c_{GB}$ is maintained one obtains by solving the diffusion law:

$$c(x,x,x,t,c_{\mathrm{GB}}) - c_{\mathrm{GB}} =$$

$$c_{\mathrm{GB}} \prod_{i=1}^{3} \left\{ \sum_{m=0}^{\infty} \frac{4}{\pi(2m+1)} \sin\left[\frac{(2m+1)\pi x_i}{l_i}\right] \exp\left[-Dt\frac{\pi(2m+1)^2}{l_i^2}\right] \right\},$$

$$m = 0, 1, 2, \ldots,$$

from which equation depth profiles may be easily calculated by integration along $x_1, x_2$. Typical shapes of $c(x_3)$ are shown in Fig. 117. From such profiles $D$ can be estimated provided $l_1, l_2$ are determined at least approximately. It is, however, easier to calculate the mean diffusivity (mixing rates) from the amount of material that traversed the interface, as will be shown later.

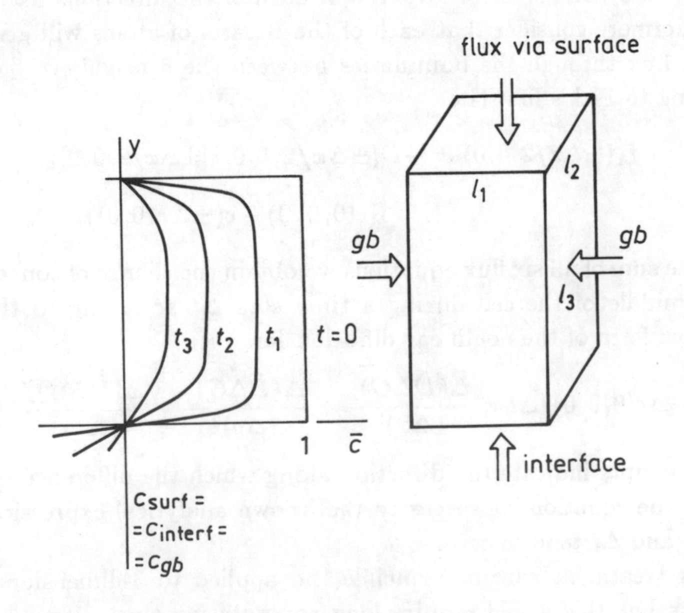


Fig. 117. Short circuit diffusion.

### 9.4.3. *Numerical approaches to GB-supported diffusion in thin films*

Thin films are ideally suited for numerical solution of complex diffusion problems on the basis of finite differences of the diffusion law. The finite volume of a thin film allows for a reasonable limit of the amount of discrete volume elements and hence reasonable computation time. The treatment is simple. Diffusion fluxes according to the actual differences of concentrations in different volume elements are summed up in discrete time steps, thus changing the concentration after each time step. This numerical treatment is directly due to the real physical process.

Let us consider a continuum of a given shape and a distribution of two components A and B according to $c_A = c(x,y,z)_{t=0}$. The diffusivity of each component is represented by an interdiffusion parameter $\tilde{D}[x,y,z,c(x,y,z)]$. This continuum is thought to be an assembly of equal-sized cubes with an edge length $\Delta x = \Delta y = \Delta z$. We choose an arbitrary cell placed with its middle at the origin of the coordinate system $\{x,y,z\}$. With respect to random walk of the atoms in the absence of a driving force, we may divide them into 1/6 in each of the directions $\pm x, y, z$. Let us furthermore consider that each of the 6 parts of atoms will give rise to a mean flux through the boundaries between the 6 neighboring elements according to Fick's first law

$$j_A(\pm\Delta x/2,0,0) = -\tilde{D}(\pm\Delta x/2,0,0,c[\pm\Delta x/2,0,0]) \{c(0,0,0) - c(\pm\Delta x,0,0)\}\Delta x \ .$$

From the sum of all six flux equations we obtain the change of concentration in the middle of the cell during a time step $\Delta t$ according to the finite-difference form of the nonlinear diffusion law

$$\Delta c(0,0,0)/\Delta t = \frac{\Delta(\tilde{D}\Delta C)_x}{(\Delta x)^2} + \frac{\Delta(\tilde{D}\Delta C)_y}{(\Delta y)^2} + \frac{\Delta(\tilde{D}\Delta C)_z}{(\Delta z)^2} \ .$$

The subscripts indicate the direction along which the difference has to be taken. The equation converges to the known analytical expression if $\Delta x$, $\Delta y$, $\Delta z$ and $\Delta t$ tend to zero.

This treatment can, in principle, be applied to 3-dimensional grain-networks but this would require long computation time. Hence, instead, 2-dimensional grain structures are considered further on. Many parameters may be introduced: limited miscibility, different GB networks, moving GB's in diffusion recrystallization, unsymmetric couples, etc.

Depth profiles of concentration, the concentration spectrum and other measurable quantities may be calculated conveniently for simulated diffusion processes and subsequently exposed to the evaluation procedures as discussed before. Valuable information concerning errors is then obtained by comparing these results with the input data. In the following the same selected examples are shown. We start from a symmetric couple of components $A$ and $B$. Each single layer contains GB's periodically distributed at the same position. Then one period of the couple may be cut out since through the symmetry planes of the grains no flux of atoms exists. The spatial distribution of such a couple in an early state of diffusion is shown in Fig. 118.

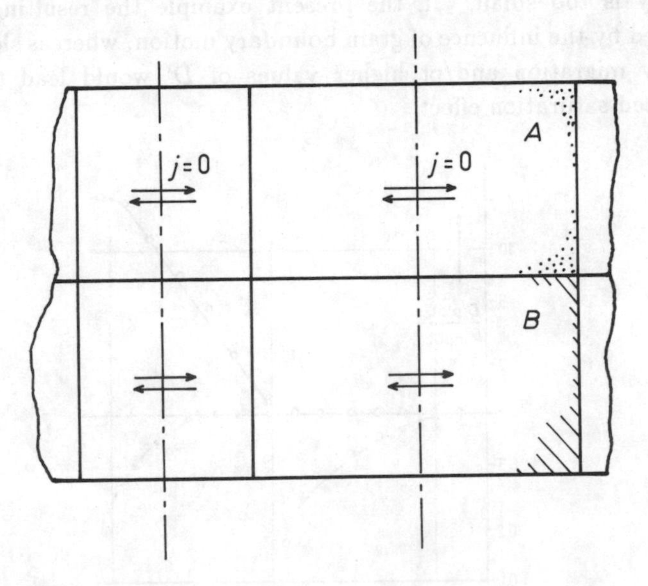


**Fig. 118.** Spatial distribution in an early state of diffusion in a symmetric couple of components A and B.

One can easily recognize the effect of the highly conducting paths where the diffusant is quickly distributed.

In many practical cases freshly deposited films start to recrystallize. The action of moving GB's on the distribution of diffusant in the host layer is quite obvious. The moving GB's enhance diffusion very strongly, which in fact has often been observed experimentally.

Whipple's analysis of computed depth profiles is as follows. On the basis of depth profile calculated by averaging the concentration across planes of the film and using the input value $D$ one may evaluate $D'$ from a Whipple plot, shown in Fig. 119. It shows that the influence of saturation, in fact appear at the diffusion state given by $\sqrt{D't}/l = 1$. Up to this state

Whipple's treatment gives excellent results whereas deviation rises quickly with proceeding $\sqrt{D't}$. The dashed curves obtained for the case of moved grain boundaries reveal deviations of $D'_{\text{plot}}$ towards smaller values. The discrepancies increase with the migration distance. Evidently the moving grain boundary crosses regions where the diffusant is not yet significantly present. The grain boundary depleted in this way has to be replenished again at each new position, thus distributing the diffusant in the interface region rather than deep into the film. Hence, the evaluated grain boundary diffusivity is too small. In the present example the resulting effect is dominated by the influence of grain boundary motion, whereas slower grain boundary migration and/or higher values of $D'$ would lead to a more pronounced saturation effect.

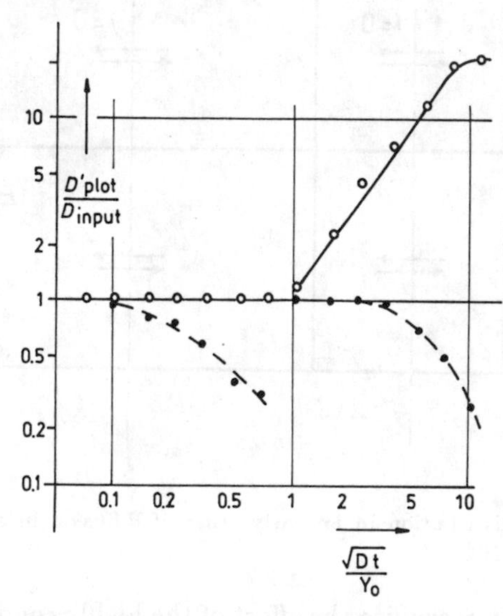


**Fig. 119.** Results of the Whipple plots of synthesized diffusion couples. Open circles: Fixed grain boundaries. Full circles: Moved grain boundaries.

Bulk diffusivity is also altered by GB motion. More diffusant is distributed next to the interface:

$$D_{\text{evaluated}} > D_{\text{input}} \ .$$

## 9.4.4. *Mean diffusion rates*

A common estimation of diffusivity $\overline{D}$ responsible for any diffusional mixing process averaged over time and space is based on the relation

$$\overline{D} = d^2/t_e \ ,$$

where $d$ is the couple thickness and $t_e$ the time of complete mixing. Since there is always a considerable uncertainty in finding $t_e$, it is desirable to assign a mean diffusivity $D_{eff}$ to certain periods during the process itself. $D_{eff}$ defined in such a way may also give information about the possible time dependence of the mixing rates. However, if it should correspond to intermediate stages of mixing, we would have to introduce a proper description of such stages first.

Independent of the actual local distribution of the concentration, which is fairly inhomogeneous in case of GB-supported diffusion, the state of mixing can be defined by the amount of diffusant (material $A$ or $B$) which has passed the original couple interface. This amount is expressed by the mean concentration $\overline{c}_B$ in the former component layer $A$ and complementarily $\overline{c}_B$ in layer $B$. Mass conservation must maintain the total concentration $c_{OA}$:

$$c_{OA}(d_A + d_B) = \overline{c}_A d_B + (1 - \overline{c}_B) d_A \, ,$$

where $d_A$ and $d_B$ denote the layer thicknesses.

In order to state the effective diffusion coefficient $D_{eff}$, we assume a diffusion couple of equal dimensions $d_A$, $D_B$ and a uniform diffusivity $D_{vol}$. In such a couple diffusion would take place as a one-dimensional process leading to the known depth distribution of the concentration expressed for instance in a Fourier form

$$c_A(z,t) = \sum_{n=1}^{\infty} a_n \cos\left(\frac{n\pi z}{d_A + d_B}\right) \exp\left[-Dt\left(\frac{n}{d_A + d_B}\right)^2\right]$$

The effective diffusivity $D_{eff}$ may now be defined as that value $D$ of this imaginary couple which would give rise to the same total mass flux through the interface within the same time interval and therefore to the same amount of materials $A$, $B$ in the opposite layers.

This value $D = D_{eff}$ corresponds uniquely to one particular pair of mean concentrations $\overline{c}_A$, $\overline{c}_B$, in the partial layers of the couple:

$$\overline{c}_A = \frac{d_A + d_B}{d_A} \sum_{n=1}^{\infty} \frac{a_n}{n\pi} \exp\left[-Dt\left(\frac{n}{d_A + d_B}\right)^2\right] \sin\left(\frac{n\pi d_A}{d_A + d_B}\right)$$

The solution of $\overline{c}_A = \overline{c}_A(Dt)$ for $Dt$, i.e., the evaluation of $D_{eff}$, is practically carried out by the comparison of $\overline{c}_A$ calculated for the synthetic couple as a

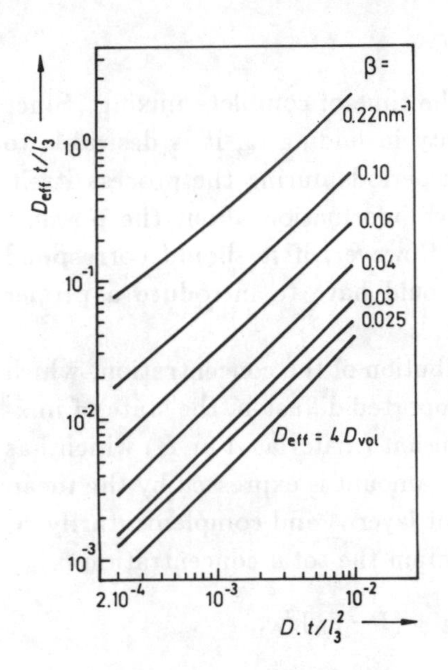


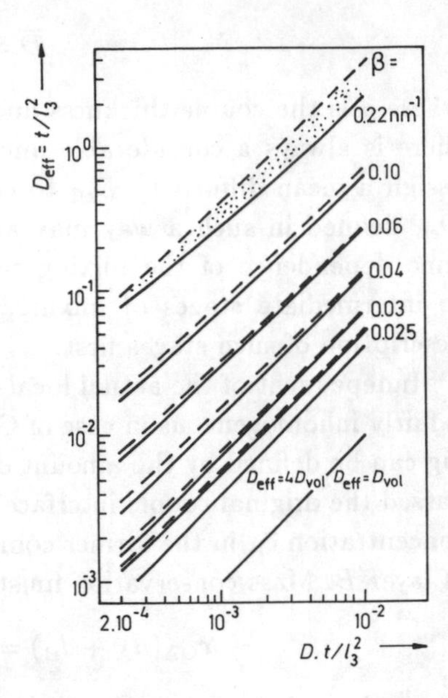


**Fig. 120. Left: Effective diffusion coefficient ($D_{\text{eff}}$) of a symmetrical couple of thickness $l_3$ as a function of the volume diffusion coefficient ($D_{\text{vol}}$) plotted in normalized values. Ideal short-circuiting is assumed along the grain surfaces. Grain shape: Prisms with square bases ($l_1 = l_2$) and height $l_3 = 100$ nm. Right: Comparison between the effective diffusivity of prismatic grains (solid lines) and slab-shaped grains (dashed lines). Ideal short-circuiting along the grain surfaces.**

function of $Dt$ with the measured actual values $\overline{c}_{\text{A}}$ (or $\overline{c}_{\text{B}}$). These values can be determined directly from experimental results. Many techniques supply the depth profiles of concentration $c_{\text{A}}(z)$, of which the mean concentration can be found by

$$\overline{c}_{\text{A}} = \int_0^d c_{\text{A}}(z)dz \ .$$

On the other hand, diffraction techniques enable the measurement of the concentration spectrum $H(c)$ given as the volume fraction with certain concentration:

$$H(c) = (1/V)dV(c)/dc$$

where $V$ is the film volume. Observe that in each partial layer the concentration $c_i$ in the interface, the mean value $\overline{c}_{\text{A}}$, $\overline{c}_{\text{B}}$, is found according to

$$\overline{c}_{\text{A}} = \int_0^{c_i} cH(c)dc \ .$$

The concentration in the interface can be roughly approximated by the total concentration, especially for later diffusion states. The accurate value $c_i$, however, is found according to

$$\int_0^{c_i} H(c)dc = V_{\mathrm{A\ or\ B}}$$

with $V_{(\mathrm{A\ or\ B})}$ denoting the volume of the partial layer under consideration. Investigation of thin film diffusion based on the measurement of only one quantity, as for instance the film resistance or the decay of X-ray satellite interferences, needs some supplementary assumptions for the distribution of concentration which directly leads to the mean concentrations and hence to $D_{\mathrm{eff}}$. As an example of the effect of a stationary and moving GB network on the mean rate of mixing, the curves calculated either for ideal short circuiting at GB's and surfaces of a symmetric couple of square prismatic grains or for a two-dimensional array of GB's are shown in Fig. 120 and 121. The GB density is described by the specific GB area. $\beta$ is the ratio of grain surface to grain volume. The film thickness in that case is assumed to be 100 nm. In the figures $D_{\mathrm{eff}}t/l_3{}^2$ is plotted against $Dt/l_3{}^2$ for different specific grain surfaces. As expected, we obtain increasing mean diffusivities for increasing $D$. If the grain shape changes to a pronounced columnar structure, i.e., $l_1/l_3 \leq 0.5$ or $\beta \leq 0.1\ \mathrm{nm}^{-1}$, the mean diffusivity changes by a factor of about 4 when one halves $l_2$ or doubles $l_3$. In this case the flux through the mantle area may be ignored. Hence, halving the lateral prism diameter means doubling the mantle surface in the original volume, as well as doubling the concentration gradient perpendicular to the mantle, which together increase the flux per film volume and consequently the speed of mixing by a factor of 4. The same result would be obtained by the following consideration. Doubling of the column length, and hence of the film thickness, would not alter the atomic flux per unit volume since the mantle area as well as the film volume is doubled. The speed of mixing is then independent of the film thickness, but should decrease in the one-dimensional reference couple. Therefore, the effective diffusivity is likewise enhanced by a factor of 4. This behavior has in fact been observed experimentally.

Turning now to disc-shaped grains, i.e., $\beta > 0.06\ \mathrm{nm}^{-1}$ or $l_1/l_3 > 1$, the values $D_{\mathrm{eff}}$ decrease with decreasing $\beta$ and should finally converge towards $4D_{\mathrm{vol}}$ since, even if GB, distances approach infinity, in our model constant concentration is maintained at all grain surfaces and hence also at the free surface. Therefore the diffusant still penetrates from the base (couple

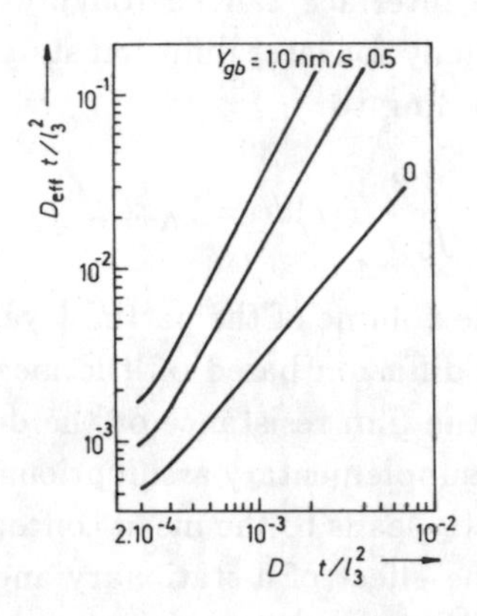


**Fig. 121. Effective diffusivity of a sample with steady-state grain boundary network ($v_{GB} = 0$) and of a sample with moving grain boundaries ($v_{GB} = 0.5, 1.0$ nm/s)**

interface) as well as from the top (film surface) into the grain. Any flux through the mantle can be neglected. This shows clearly the limited validity of the assumption of infinite GB and surface diffusivity. In practice the surface will not be adequately supplied with the diffusant through a GB network of low density. In the case of complete absence of GB, $D_{eff}$ is obviously identical to $D_{vol}$.

As one can see, the mean diffusivity is strongly dependent on the grain surface. The more pronounced the columnar shape (high GB density or high $\beta$) the higher is $D_{eff}$, which may exceed $D$ by several orders of magnitude.

9.4.4.1. *Diffusion into moving GB* (*Two-dimensional model*)

The mutual influence of diffusion and recrystallization has been discussed on the basis of experimental and computed results. Here we wish to point out the effect of diffusion via moving GB's on the effective diffusivity. The GB network of $\beta = 0.025$ nm$^{-1}$ and the diffusion data used in the above section were used and in addition the migration of the GB's with a constant speed of 0.5 and 1.0 nm/s was assumed. In each new position the grain boundary is replenished with the diffusant thus leaving a track of diffused material in the grain volume which it had traversed. Hence the material transfer is not directly governed by diffusion but rather by the speed of the GB migration, if diffusion into the GB is sufficiently fast. This is the case in our examples. We notice a considerably higher mean diffusivity. From the slope of $D_{eff}$ we deduce that $D_{eff}$ even rises with time. This obviously holds only as long as the GB's move without mutual annihilation. When the GB density is reduced, we should expect a decrease of $D_{\text{eff}}$. Here we only wish to demonstrate the enhancing effect of GB's on the mean diffusivity while they move.

### 9.4.5. *Diffusion in systems with miscibility gaps and intermediate phases*

Systems with partially miscible partners do not differ too much in diffusion behavior from miscible components as long as no intermediate phases are formed. The only difference is that we have a boundary condition requiring that for all surfaces the concentration of the edges of the gap has to be maintained. Practically, however, the solubility of the diffusant in the GB's may be also limited and in thin films the gap edges may differ from equilibrium values, see Fig. 122.

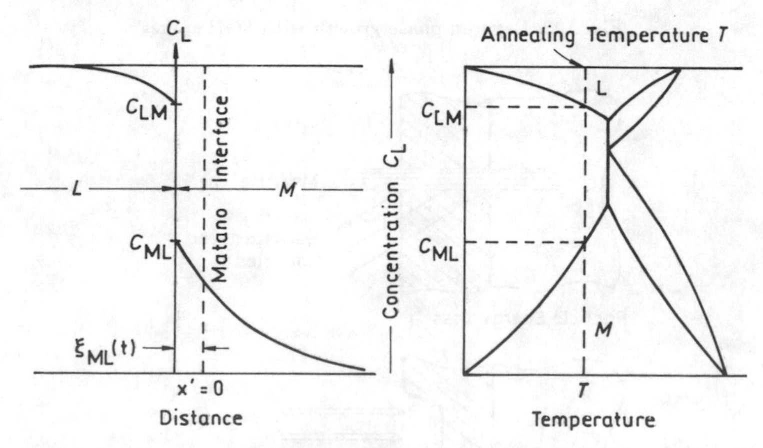


Fig. 122. Interdiffusion of partially miscible metals (single-crystal films). At the temperature of anneal $T$, the limit of solubility of $M$ in $L$ is $C_{\text{ML}}$ and the solubility of $L$ in $M$ is $1 - C_{\text{LM}}$.

In phase-forming systems the formation and further growth of the compounds are complex problems. Usually not all thermodynamically possible compounds are simultaneously formed in thin films. The type of diffusion is mainly ruled by how and where the diffusant crosses the phase just formed. Different processes that have been found experimentally are to be discussed later. Again, in thin films one has to be aware that non-equilibrium diffusion may give rise to completely unexpected behaviors, e.g., extended miscibility, formation of non-equilibrium phases etc, see Fig. 123.

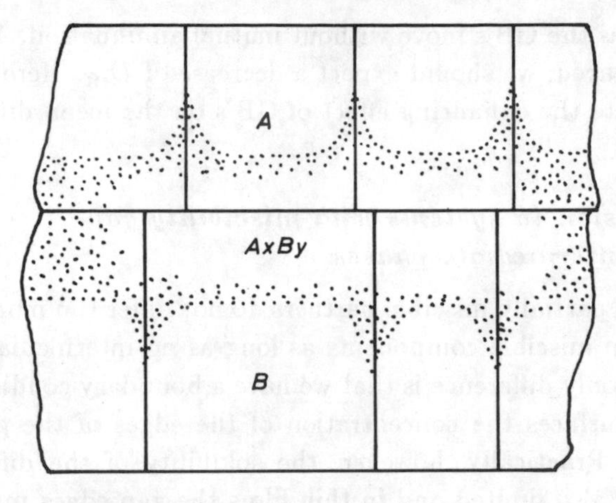


**Fig. 123. Layered phase growth with "GB spikes".**

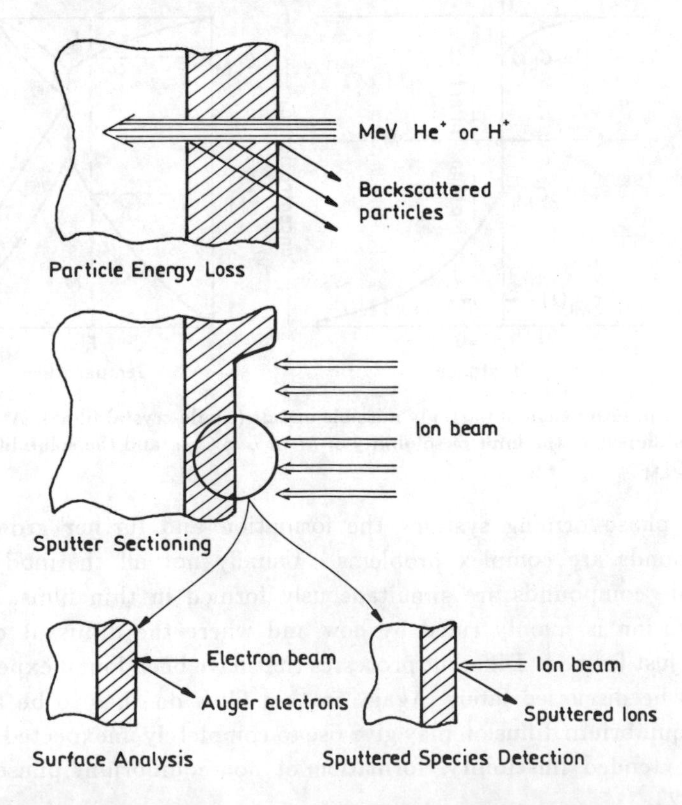


**Fig. 124. Principles of various surface analytical techniques.**

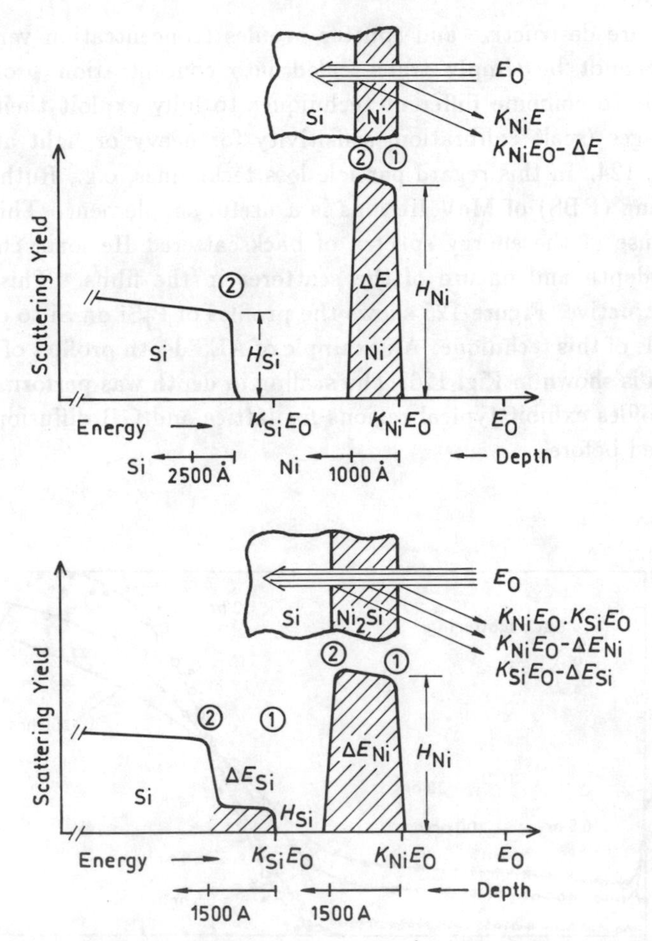


Fig. 125. Schematic backscattering spectra for MeV $^+$He ions incident on Ni film on Si (top) and after reaction to form $Ni_2Si$ (bottom). Depth scales are indicated below the energy axes. Circled numbers 1 and 2 indicate the outer surface and the interface, respectively.

## 9.5. Selected Experimental Results[20–23]

### 9.5.1. *Experimental techniques*

As already pointed out in Sec. 9.3.1, lattice and GB diffusion parameters may be obtained; however depth profiles on the scale of nonometers need special equipment favorably combining sputter profiling with one of the surface analytical techniques. Widely used is Auger electron spectroscopy (AES) or secondary ion mass spectrometry (SIMS). Such methods, of course, are destructive and sputter profiles (concentration versus sputter time) cannot be simply trans-scaled into concentration profiles. It is desirable to combine different techniques to fully exploit their particular advantages (scale calibration, sensitivity for heavy or light atoms, etc.), see Fig. 124. In this regard particle loss techniques, e.g., Rutherford back scattering (RBS) of MeV He ions is a useful supplement. This technique makes use of the energy spectra of backscattered He ions, characteristic of the depth and nature of the scatterer in the films. This method is nondestructive. Figure 125 shows the profiles of PtSi on Si to envisage the principle of this technique. An example of AES depth profiles of Pd diffused into Au is shown in Fig. 126. The scaling in depth was performed by RBS. The profiles exhibit typical regions for lattice and GB diffusion regions as discussed before.

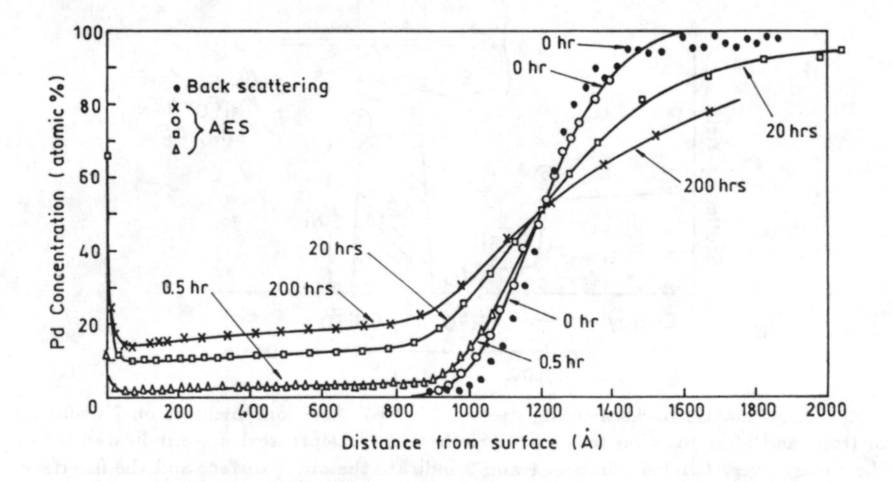


Fig. 126. Palladium concentration profiles in Pd-Au couples.

In principle any concentration-dependent property may be used as indicator for diffusional mixing. For example, X-ray, optical and resistometric methods are by far simpler, though generally at the expense of straight forwardness and sensitivity. On the other hand, as non-destructive methods, their striking advantage is that they are suitable for following the diffusion process in one and the same sample.

# CHAPTER 10

# MECHANICAL PROPERTIES OF THIN FILMS[24–27]

## 10.1. General

The field of mechanical applications of films, e.g., for reduction of friction, wear and abrasion, is prospering since some years ago and special methods for the production of such films have developed rapidly, hand in hand with the new techniques for determination of their mechanical properties. Two of the latter are not only interesting in films for mechanical purposes but obviously for any practically used layers.

Adhesion to the substrate and internal stresses may strongly influence the physical properties.

In the following sections we shall give an overview on the mechanical properties, mainly on the basis of the film microstructure and the parameters influencing them. New preparation techniques and methods for the characterization are discussed. Special applications and materials will be mentioned.

## 10.2. Micromorphology of Evaporated and Sputtered Deposits[28]

### 10.2.1. *Evolution of the microstructure*

As pointed out before the general steps in film formation are adsorption, nucleation, growth of nuclei and coalescence and growth of the continuous film. It is clear that those parameters specifically influencing each step of formation will determine the development of the final microstructure. Hence the dominant variables are the nature of substrate, substrate temperature, deposition rate, angle of particle incidence, thickness, nature, state and amount of further impinging particles. Ion bombardment has been found to have a large influence on the interface structure and thus on the adhesion of coatings. The mechanism involved may be an increase in the density of nucleation sites, and therefore a reduction in the void formation during the initial island growth.

The nature of the subsequent growth process is such that the structure, and therefore the properties, of vacuum-deposited coatings in general and sputtered coatings in particular are determined largely by the selection processes that are made evolutionary by the way how the state of the coating surface changes as the coating grows.

Any process that causes a systematic non-uniformity in the arriving coating atom flux over the substrate surface can have a drastic effect on the evolutionary growth process. Shadowing, a simple geometric interaction between the roughness of the growing surface and the line-of-sight directions of the arriving coating atoms, provides such an effect. For many pure metals the adatom binding energies and the activation energies for both surface and bulk diffusion are related and proportional to the melting point. Furthermore, the shadowing effect can be compensated by surface diffusion, i.e., if the surface diffusion is large enough, the point of arrival of a coating atom loses its significance. Thus various basic processes including shadowing can be expected to dominate over different ranges of $T/T_m$ and to manifest themselves as differences in the resulting coating structures, where $T$ is the substrate temperature and $T_m$ the melting point of coating material. Such is the basis for the structure zone models described below.

For the melting point of a particular coating material and the working pressure of sputter gas four characteristic structures are developed. This diagram holds as well for evaporated films. The transition zone may be rather suppressed in the case of evaporated pure metals. The evolution of the structural morphology is as follows.

At low temperatures, the surface mobility of the adatoms is reduced and the structure grows as tapered crystallites from a limited number of nuclei. It is not a structure of full density but contains longitudinal porosity of the order of a few hundred angstroms wide between the tapered crystallites. It also contains a high dislocation density and has a high level of residual stress. Let us call this structure Zone 1.

As the substrate temperature increases, the surface mobility increases and the structural morphology first transforms to that of Zone $T$, i.e., tightly packed fibrous grains with weak GB's, and then to a columnar morphology of full density, called Zone 2.

The size of the columnar grains increases as the condensation temperature increases. Finally, at still higher temperature the structure shows an equiaxed grain morphology, the Zone 3. It was shown that the evolution of Zone 2 is ruled by surface self diffusion and that of Zone 3 by volume self diffusion.

The transition temperatures may vary significantly from those stated above and in general shift to higher temperatures as the gas pressure in the synthesis process increases.

It should be emphasized that:

(1) The transition from one zone to the next is not abrupt but smooth. Hence the transition temperatures should not be considered as absolute but as guidelines only.
(2) All the zones are not found in all deposits. For example, Zone $T$ is not prominent in pure metals, but becomes more pronounced in complex alloys, compounds, or deposits produced at higher gas pressures. Zone 3 is not seen very often in materials with high melting points.

Most thick deposits exhibit a strong preferred orientation (fiber texture) at low deposition temperatures and tend towards a more random orientation with increasing deposition temperature. For example, observations showed the evolution of a large grained columnar morphology in a Be deposit from a much larger number of fine grains which were originally nucleated on the substrate. As the growth proceeds, only those grains with a preferred growth direction survive, presumably due to the minimization of the surface energy.

The texture of evaporated deposits is in general dependent on the deposition temperature. At low deposition temperatures, a strong preferred orientation is generally observed: [211] in Fe, [220] in TiC, and [0002] in Ti, etc. As deposition temperature increases, the texture tends to become more random. In the case of Be, the texture changes to a [1120] orientation at high deposition temperatures. The presence of a gas tends to shift the preferred orientation to higher index planes. For Ag, increasing the substrate bias changes the preferred orientation from [111] to [200] and back to [111].

Further influences on the evolution of the microstructure of a film are substrate roughness, oblique deposition, inert gas effects, and ion bombardment-bias sputtering.

The surface roughness of practical substrates often exists simultaneously on several size scales. The resultant Zone 1 coating structure will then exhibit superimposed arrays of shadow growth boundaries, each associated with a size scale of substrate roughness. As $T/T_m$ is increased, successively more severe shadow boundaries are overcome by diffusion. Severe surface irregularities can lead to boundaries of Zone 1 type that persist to $T/T_m$ ranges of Zone 2 and 3. Such boundaries have been referred to as columnar and linear defects.

Substrate surface irregularities such as inclusions or debris particles that fall on the substrate or growing coating can cause preferential nucleation and runaway growth. These nodular or flake defects can form at both low and high $T/T_m$.

As a general rule, if a critical coating is required for high performance applications, one of the most effective steps that can be taken is to provide a smooth homogeneous substrate.

Oblique deposition and substrate surface roughness are complementary in producing shadow-induced Zone 1 boundaries. Therefore it is difficult to achieve uniform dense microstructures in coatings deposited over complex shaped substrates by vacuum deposition at moderate $T/T_m$. Zone 1 crystals tend to point in the direction of the coating flux.

An elevated inert gas pressure at low $T/T_m$ causes the Zone 1 boundaries to become opener. This effect has been observed in coatings deposited by magnetron sputtering (no plasma bombardment) and by evaporation. Planar diode sputtering sources must necessarily operate at high working gas pressures. However, the resulting coating structures are not as voided as would be expected, because the associated plasma bombardment tends to suppress the deleterious effects of the working gas.

Intense substrate ion bombardment during deposition can suppress the development of open Zone 1 structures at low $T/T_m$. The deposits have a structure similar to the Zone $T$ type. Ion bombardment on uncooled substrates yields typical high $T/T_m$ structures.

The ion bombardment may suppress the Zone 1 structure by creating nucleation sites for arriving coating atoms or by eroding surface roughness peaks and redistributing material into valleys.

### 10.2.2. *Internal stresses*

Virtually all vacuum deposits are in a state of stress. The total stress is composed of a thermal stress due to the difference in the thermal expansion coefficient of the coating and substrate materials, and an intrinsic stress due to the accumulating effect of atomic forces generated throughout the coating volume by atoms which are out of position with respect to the minimum in the interatomic force fields.

For materials of low melting point, the deposition conditions will generally involve sufficiently high values of $T/T_m$ so that the intrinsic stresses are significantly reduced by recovery during the coating growth. Thermal stresses are therefore of primary importance for such materials. Stress relief can occur as the coatings are brought to room temperature or annealed following deposition. The resultant material flow can lead to formation of hillocks or holes, depending on whether the temperature change places the coating into compression or into tension. Thus extensive hillock growth has been reported in Al, Au and Pb films at elevated temperatures.

Materials of high melting points are generally deposited at sufficiently low $T/T_m$ ($< 0.25$) so that the intrinsic stresses dominate over the thermal stresses. For thin films ($< 500$ nm thick) the intrinsic stresses are generally constant throughout the coating thickness. They are typically tensile for evaporated and often compressive for sputtered metal coatings. Magnitude can be near the yield strength for the coating materials and can reach $10^9$ N/m$^2$ for refractory metals such as molybdenum and tungsten.

The interfacial bond must withstand the shear forces associated with the accumulated intrinsic stresses throughout the coating as well as the thermal stresses. Since the intrinsic stress contribution increases with coating thickness, it can be the cause of premature interface cracking and poor results in adhesion test for coatings with thickness exceeding a critical value as low as 100 nm.

#### 10.2.2.1. *Model for growth-induced stress*

A somewhat simplified model of film formation is assumed in which the nuclei grow first as hemispheres until they touch and subsequently as columns with densely packed boundaries. The interatomic forces at the boundaries tend to close any existing gap, with the result that the neighboring crystallites are strained in tension.

During the very first nucleation stage, each nucleus is strained only by its surface tension, but can migrate quite freely over the substrate surface until it reaches a radius where the bonds of the iron atoms to the surface freeze due to the strains imposed by surface tension. This anchoring of the crystallites occurs for iron at a radius of about 1 nm, and it induces the first measurable strain into the substrate. The contraction of a crystallite of this size due to surface stress is of the order of 1%.

The next phase represented by the further growth of the anchored but still isolated nuclei is characterized by an expansion of the crystallite lattice due to the decreasing influence of the surface tension. Thus, a force versus thickness curve determined by a bending plate experiment exhibits no stress up to an island size of the order of 1 nm, but has a region of increasing compressional force up to the point where the islands touch and grow together. Such behavior is indicated schematically in Fig. 127. As the grains begin to form a continuous layer, many unclosed gaps still exist between the crystallites.

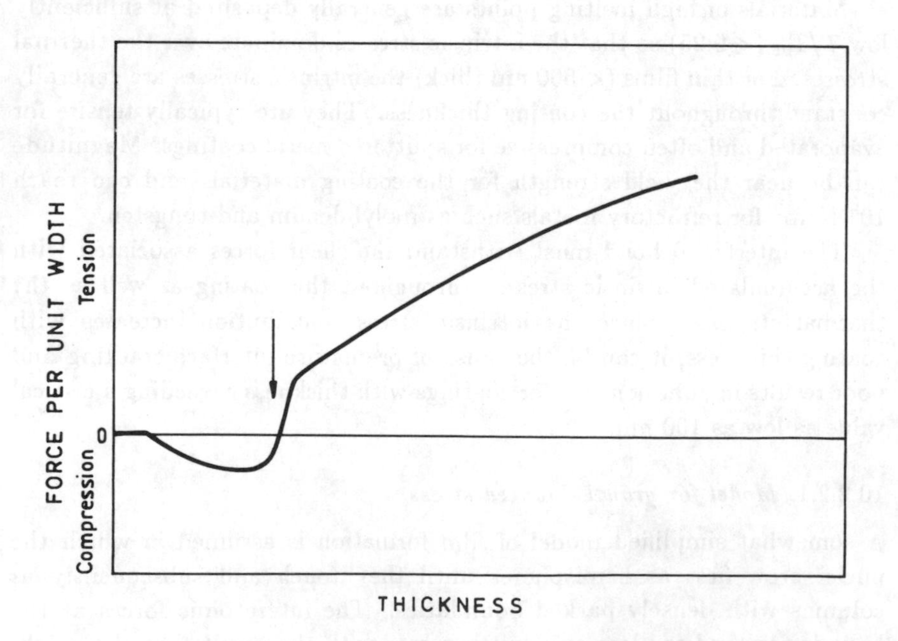


Fig. 127. Schematic representation of force per unit width following composite surface tension-crystallite mismatch relaxation model. The thickness indicated by the arrow corresponds to the film becoming continuous.

A detailed analysis shows that the closing of these gaps results in a sharply increasing tensional stress. After the film has reached a sufficient thickness so that it has passed through this transitional stage and reaches an average strain value equal to the average atomic relaxation distance divided by the average crystallite dimension. An average interatomic force which follows Hooke's law for small displacements and decreases exponentially for large displacements is assumed. For the case of increasing crystal size during film growth, the stress increases more slowly than linear, as indicated in the figure.

10.2.2.2. ***Further influence not accounted for in the model***

Stresses were found to be particularly sensitive to the working gas species and pressure and to the apparatus geometry and angle of incidence of the coating flux relative to the substrate surface. Figure 128 shows the influence of the argon working-gas pressure on the interface force per unit length (integrated stress) that develops in coatings, approximately 200 nm thick, of several materials deposited at normal incidence on substrates at near-room temperatures. The general behavior has been observed for more than ten metals ranging in atomic mass from that of Al to W and for amorphous Si.

## 10.3. Determination of Mechanical Properties in Thin Films

### 10.3.1. *Internal stress*

The mean stress across the film is compensated by an equivalent counter-force from the substrate. If the substrate is chosen thin enough the film stress causes the substrate to bend. Bending cantilevers or membranes are therefore a proper tool to measure and monitor internal stress.

The deflection of a circular plate has been used for stress measurements. The change in the optical fringe system between the plate and an optically flat plate is used here to measure the deflection of the plate. Because of the limited flatness of available substrates, the substrate profile is measured after the film has been dissolved and then used as a reference profile. Fused quartz substrates are generally flatter than glass, but on the other hand have more severe thickness gradients. The fringe technique is illustrated in Fig. 129. The circular plate offers the possibility of observing stress anisotropy. Anisotropy has also been observed by using two orthogonal cantilever beams.

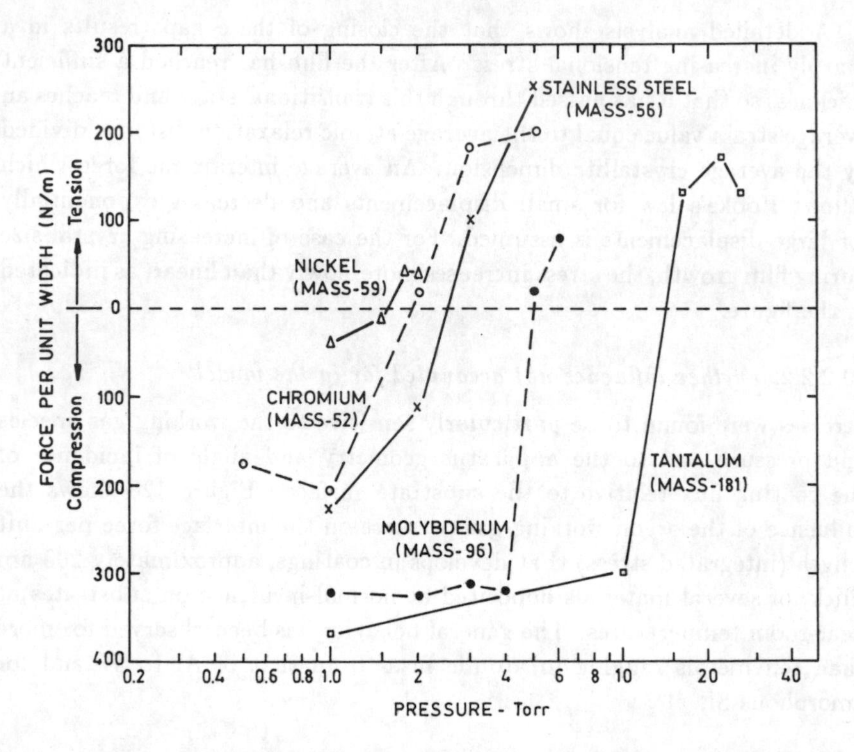


**Fig. 128. Force per unit width produced at substrate by internal stress in coatings deposited at various argon pressures using cylindrical-post magnetron sputtering sources.**

In both the beam and plate methods, it is the elastic theory which yields the film stress as a function of the measured deflection. It is assumed that the film strains the substrate, which bends until equilibrium is reached, and that the film-substrate bond is strong enough to suppress slippage, a condition that seems to be fulfilled in practice. Since the film stress is computed from the observed substrate strain only, it is apparent that the elastic constants of the film are not taken into consideration in this approximation. The theory of plates yields for an isotropic stress in equilibrium with the resultant strain,

$$\sigma = \frac{Ed^2}{6(1-\nu)rt},$$

where $\sigma$ is the stress in the film, $E$ the Young's modulus of the substrate, $d$ the substrate thickness, $t$ the film thickness and $r$ the radius of curvature of the bent strip, and $\nu$ the Poisson ratio of the substrate.

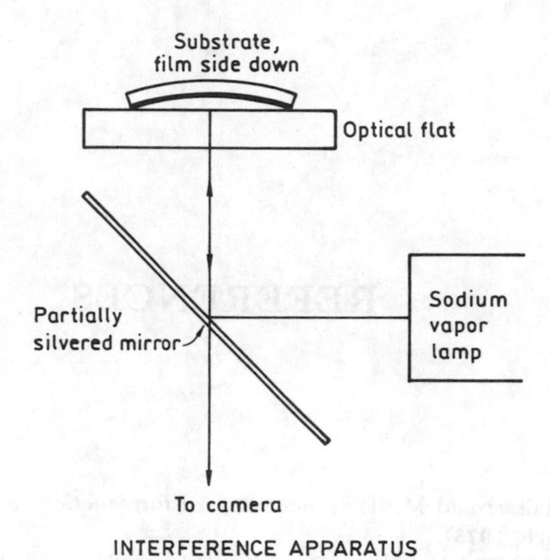


**Fig. 129. The fringe technique for measuring the internal stresses.**

Stresses may also be deduced from the measurement of changes in the lattice parameters. X-ray techniques are preferable to electron diffraction because of their higher resolution, owing to the larger Bragg angles. Kinbara and Haraki have measured the lattice constant perpendicular to the film plane by a diffractometer. The strain is measured directly and the isotropic stress is then computed from

$$\sigma = \frac{E(a - a_0)}{2\nu_f a_0},$$

where $a_0$ and $a$ are the lattice constants of the bulk material and the strained film respectively, $\nu_f$ the Poisson ratio of film. A smaller lattice constant perpendicular to the film corresponds to a tensile stress while negative values of $\sigma$ indicate compression.

The stress in the film can also be obtained from a measurement of the lattice constant in the film plane, although the geometry is here more suitable for electron diffraction. The stress is given in this case by

$$\sigma = \frac{E(a - a_0)}{(1 - \nu_f)a_0}.$$

# REFERENCES

1. C. I. Maissel and M. H. Francombe, *An Introduction to Thin Films* (Gordon & Breach, 1973).
2. R. J. Hill, *Physical Vapor Deposition* (Airco Temescal, 1976).
3. O. S. Heavens, *Thin Film Physics* (Methuen, 1973).
4. B. Chapman and J. Anderson, *Science and Technology of Surface Coatings* (Academic, 1974).
5. J. W. Mathews, *Epitaxial Growth, Part A, B* (Academic, 1975).
6. W. T. Tsang, *Appl. Phys. Lett.* **45**, 1234 (1984).
7. Z. Knittl, *Optics of Thin Films* (John Wiley, 1976).
8. H. Ebel, A. Wagendristel and H. Judtmann, *Z. Naturforschung*, **23a**, 1863 (1968).
9. T. J. Coutts, *Electrical Conduction in Thin Metal FIlms* (Elsevier, 1974).
10. C. Dupuy and A. Cachard, *Physics of Nonmetallic Thin Films* (Plenum, 1974).
11. R. J. Coutts, *Active and Passive Thin Film Devices* (Academic, 1978).
12. D. H. Navon, *Semiconductor Microdevices and Materials* (CBS College Publishing Co., 1986).
13. H. W. Lam and M. J. Thompson, *Comparison of Thin Film Transistor and SOI Technologies, MRS Symposia Proceedings Vol.* 33 (Elsevier, 1984).
14. P. N. Butcher, N. H. March and M. P. Tosi, *Crystalline Semiconductor Materials and Devices* (Plenum, 1986).
15. H. Hoffmann, *Thin Films: Trends and New Applications, Vol. 2* (Elsevier, 1989).
16. A. Wagendristel, *Appl. Phys.* **7**, 175 (1975).
17. A. Wagendristel, *Z. Naturf.* **30a**, 1648 (1975).
18. A. Wagendristel, *Phys. Stat. Sol.* **A13**, 131 (1975).
19. A. Wagendristel, E. Tschegg, E. Semerad and H. Bangert, *Appl. Phys.* **10**, 237 (1976).

20. Maria F. Ebel and A. Wagendristel, *Proc. 7th International Vacuum Congress and 3rd International Conference of Solid Surfaces*, p. 2149 (1977).
21. P. Schattschneider and A. Wagendristel, *Z. Naturf.* **33a**, 693 (1978).
22. A. Wagendristel, H. Bangert and W. Tonsern, *Surf. Sci.* **86**, 68 (1979).
23. A. Wagendristel, H. Schurz, E. Ehrmann-Falkenau and H. Bangert, *J. Appl. Phys.* **51**, 4808 (1980).
24. Ludmila Eckertova, *Physics of Thin Films* (Plenum, 1977).
25. John D. Dow and Ivan K. Schuller, *Interfaces, superlattices and thin films, MRS Symposia Proceedings, Vol. 77*, 1987.
26. H. Bangert and A. Wagendristel, *Rev. Sci. Instr.* **56**, 1586 (1985).
27. H. Bangert and A. Wagendristel, *J. Vacuum Sci. Techn.* **A4**, 2956 (1986).
28. J. E. Sundgren, *Thin Solid Films* **128**, 21 (1985).

# INDEX